Introduction to Engineering
A Starter's Guide With Hands-On Digital Multimedia and Robotics Explorations

Introduction to Engineering: A Starter's Guide With Hands-On Digital Multimedia and Robotics Explorations
Lina J. Karam and Naji Mounsef
www.morganclaypool.com

ISBN: 9781598297621   paperback
ISBN: 9781598297638   ebook

DOI: 10.2200/S00140ED1V01Y200806ENG007

A Publication in the Morgan & Claypool Publishers series

SYNTHESIS LECTURES ON ENGINEERING #7

Lecture #7

Series ISSN
ISSN: 1939-5221   print
ISSN: 1939-523X   electronic

Introduction to Engineering

*A Starter's Guide With Hands-On Digital
Multimedia and Robotics Explorations*

Lina J. Karam and Naji Mounsef
Arizona State University

SYNTHESIS LECTURES ON ENGINEERING #7

MORGAN & CLAYPOOL PUBLISHERS

ABSTRACT

This lecture provides a hands-on glimpse of the field of electrical and computer engineering. The broad range of hands-on applications utilize LabVIEW and the NI-SPEEDY-33 hardware to explore concepts such as basic computer input and output, basic robotic principals, and introductory signal processing and communication concepts such as signal generation, modulation, music, speech, and audio and image/video processing. These principals and technologies are introduced in a very practical way and are fundamental to many of the electronic and computerized devices we use today. Some examples include audio level meter and audio effects, music synthesizer, real-time autonomous robot, image and video analysis, and DTMF modulation found in touch-tone telephone systems.

KEYWORDS

Digital, hands-on applications, real-time, robotics, digital signal processing, communications, control, mutimedia, digital filtering, audio, music, image, video, NI LabVIEW, NI SPEEDY-33

Foreword

This volume provides a hands-on glimpse of the field of electrical and computer engineering. The broad range of hands-on applications utilize LabVIEW and the NI-SPEEDY-33 hardware to explore concepts such as basic computer input and output, basic robotic principals, and introductory signal processing and communication concepts such as signal generation, modulation, music, speech, and audio and image/video processing. These principals and technologies are introduced in a very practical way and are fundamental to many of the electronic and computerized devices we use today. Some examples include audio level meter and audio effects, music synthesizer, real-time autonomous robot, image and video analysis, and DTMF modulation found in touch-tone telephone systems.

The provided hands-on applications can be easily expanded into longer-duration design projects. Examples of such design projects can be found at http://www.fulton.asu.edu/~karam/introeng.

The software and hardware required for the experiments in this manual are LabVIEW DSP 8.2 or newer and the NI SPEEDY 33 DSP board for Chapters 1–10 and the Vision Development Module 8.2 or newer for Chapters 11–12.

National Instruments Part Number Information

| 778010-23 | LabVIEW DSP Bundle with SPEEDY-33 (Academic Use Only) |
| 194875-01 | Robotics Daughter Module for SPEEDY-33 |

Contents

CHAPTER 1

Getting Familiar With LabVIEW and SPEEDY-33

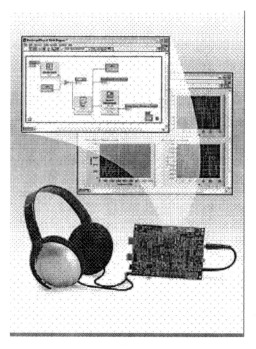

1.1 OVERVIEW

In this chapter, you will get familiar with LabVIEW's Embedded DSP Module and the SPEEDY-33 DSP board. As a starting point, you will use the basic Digital Signal Processing (DSP) capabilities in LabVIEW, such as sampling and reconstruction of continuous-time signals (A/D and D/A operations) in addition to displaying signals as a function of time.

1.2   BACKGROUND

1.2.1   What is LabVIEW?

LabVIEW stands for *Lab*oratory *V*irtual *I*nstrument *E*ngineering *W*orkbench. It is a graphical programming language from National Instruments.[1] The true power of LabVIEW lies in its ability to interface with external devices and/or internal sound cards that are installed on the PC. Therefore, it is usually used for data acquisition, instrument control, and industrial automation.

The LabVIEW DSP Module adds features to LabVIEW for signal processing applications to run on embedded DSP hardware. The National Instruments LabVIEW DSP Module can be easily interfaced with NI SPEEDY-33 and Texas Instruments DSP Starter Kits, including the TI 6713 and TI 6711 to implement and analyze DSP systems.[2]

With LabVIEW, users use graphical programming methods to learn and develop applications, and they can take advantage of the LabVIEW DSP Module to quickly implement DSP fundamentals on DSP hardware without going through the tedious task of having to write any C code, assembly, or script source.

LabVIEW programs are called virtual instruments (VIs). Each VI has two representations: a block diagram and a front panel. Each VI in turn can contain sub-VIs and other structures. Controls and indicators on the front panel allow an operator to input data into or extract data from a running virtual instrument.

The graphical code is compiled, rather than interpreted. Compilation is done on-the-fly, as the graphical code of a VI is being edited. The generated code is somewhat slower than equivalent compiled C code. However, this is considered a small price to pay for the increased productivity offered by the unique patented graphical code design system. The developed algorithms are then downloaded to the DSP board, which then runs the algorithm in a real-time environment.

You can browse different LabVIEW modules by going to the Connexions website at http://cnx.org/. Connexions is a rapidly growing collection of free scholarly materials and a powerful set of free software tools.

1.2.2   What is Speedy-33?

SPEEDY stands for *S*ignal *P*rocessing *E*ducational *E*ngineering *D*evice for *Y*outh.

The SPEEDY-33 is an easy-to-use board that contains TI's floating-point TMS320VC33 DSP. The DSP processor optimizes speed by implementing functions in hardware rather than software. It connects to a standard PC via a Universal Serial Bus (USB) host port. The VC33 DSP Education Kit from NI includes the SPEEDY-33 and the LabVIEW DSP Embedded software.

---

[1]To get more info about National Instruments, go to http://www.ni.com.

[2]To get more info about Texas Instruments, go to http://www.ti.com.

Let us take a look at the components of the SPEEDY-33 board:

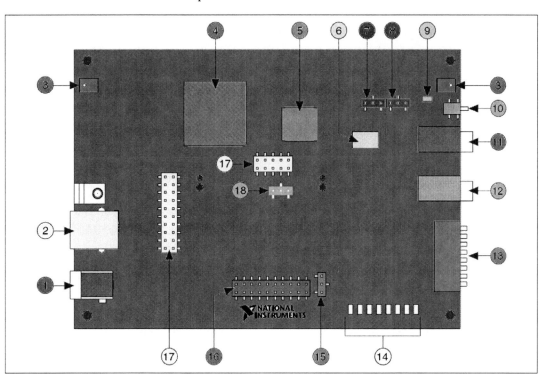

There are several things on this board that you should observe (see corresponding list numbers below). They are:

1. On the lower left end of the picture is the power port. It can supply the NI SPEEDY-33 with external power when operating the device in stand-alone mode for live demonstrations and real-world proof-of-concept applications, without being connected to the PC through USB. When power is supplied to the device, the power LED lights. The input voltage must be 9 V DC (VDC), at 500 mA, with the outside contact being ground and inner contact being positive VDC.

2. Right above, it is the connector that attaches to a USB port of your PC. LabVIEW programs and runs the DSP through this connection.

3. The two small rectangular devices at the top left and top right are onboard microphones.

4. The biggest thing on the board is the processor or DSP. Texas Instruments (TI) makes this particular DSP. TI makes many different DSPs; the "TMS320VC33" tells which chip this one is. This DSP chip has 150 written on the top. That means the processor is running at 150 MHz. This is rather slow compared to computers running at 1 to 3 GHz; however, the processing speed is more than suffice for our experiment; to process your voice, music, and video.

5. Right next to the DSP is the flash memory. Once the flash is programmed, the board can be "unplugged" from the PC, powered (9 VDC), and run by itself in a stand-alone mode for live demonstrations and real-world proof-of-concept applications.

6. One of the chips is the Audio Interface Controller. Its job is to convert the analog signal from the microphone into a digital form the processor can use. It also takes digital data from the processor and converts it into an analog form the speaker can use. This is often called an analog-to-digital converter (A/D or ADC) and a digital-to-analog converter (D/A or DAC). The 16-bit stereo audio CODEC included on the NI SPEEDY-33 allows for up to 48 kHz of dual-channel sampling on the input signal. Software components are included in the LabVIEW DSP Module to allow 8, 18, 24, 36, and 48 kHz sample rates to be used in applications.

7 and 8. The audio input level jumpers (J1, J2) control the amount of gain applied to the input audio signals. It is used to choose between Line Input (11) and On-Board Microphones (3).

9. When power is supplied to the device, the power LED lights.

10. The reset button is a small push button that manually resets the DSP and is usually used in the event of a software or hardware freeze.

11 and 12. The two connectors on the right end are Audio Stereo Input Port and Audio Stereo Output Port. They are used for attaching audio equipment. The upper one goes to the microphone. The lower one goes to the speakers.

13 and 14. There are eight lines of digital Input-Output (I/O) on the board: Input Ports are represented by eight switches (lower right) and Output Ports are represented by eight LEDs (right bottom). The digital I/O lines can be programmed with the LabVIEW DSP Module.

15. The Flash Boot Jumper controls whether the DSP will attempt to boot from the flash memory, which is the default setting. It must be enabled to run the device in stand-alone mode.

16. The simple 20-pin expansion header allows for easy interface to external hardware. The header includes power, ground, eight digital inputs, and eight digital outputs under DSP control.

17. Two connectors make up the standard expansion analog I/O connectors, which can be used for optional daughter modules.

18. The flash write enable jumper controls whether the DSP can write to the flash memory or not. This is useful in write protecting the DSP algorithm for production purposes.

1.3   SIGNAL AND TIME REPRESENTATION

Signals are the means by which we transmit information. A *signal* is defined as any physical quantity varying with time (one-dimensional signal) such as music or speech, or with space (2-D signals) such as images, or with both time and space (3-D signals) such as video and propagating waves in space. Signals exist in several types. Most of the signals in the real-world are *continuous-time* or *analog signals* which have values continuously at every value of time. To be processed by a computer, a continuous-time signal has to be first *sampled* in time into a *discrete-time signal* so that its values, at a discrete set of time instants, can be stored in computer memory locations.

The most common representation of signals and waveforms is in the time domain. For example, the following is a representation of a simple sinusoidal signal in time.

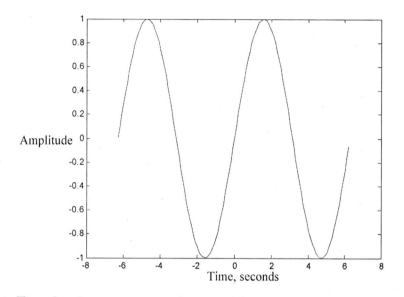

FIGURE 1.1: Time-domain representation of a sinusoidal signal.

The time-amplitude axes on which the sinusoid is shown, define the *time plane*.

1.4   STRUCTURES IN LABVIEW: WHILE LOOP, FOR LOOP, CASE

In LabVIEW, structures are represented by graphical enclosures. The graphical code enclosed by a structure is repeated or executed conditionally. There are three main structures: While Loop, For Loop, and Case.

1.4.1   While Loop

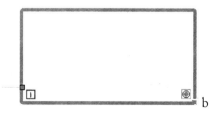

While Loop executes a subdiagram until one condition is met or the stop button is pressed. The While Loop takes in a conditional input and executes the subdiagram until the conditional terminal receives a specific value that satisfies the conditional input. The iteration terminal ⊞ (an output terminal) contains the number of completed iterations. The iteration count always starts at zero.

1.4.2   For Loop

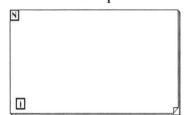

For Loop executes a subdiagram a set number of times. The value in the count terminal (input terminal), represented by , indicates how many times to repeat the subdiagram. The iteration terminal (output terminal) ⊞ contains the number of completed iterations.

The iteration count always starts at zero.

1.4.3   Case Structure

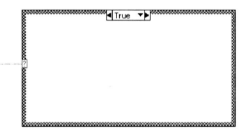

Case structure executes a subdiagram depending on the value it receives in its selector terminal ⊡. The case selector ◀ True ▼▶ shows the status being executed.

1.5   EXPERIMENT

Now that you have an idea about LabVIEW and the SPEEDY-33 board, you are ready to begin the experiment.

1.6   HARDWARE AND SOFTWARE SETUP

For all the experiments using the SPEEDY-33 board, you will have to do the following instructions at the beginning of each experiment.

- Log on to the computer
- Connect a USB cable between the USB port of the SPEEDY-33 and the USB port of your PC. The SPEEDY-33 is powered by the USB port. Make sure that the Power LED on the SPEEDY-33 is on.
- Connect the speakers to the Audio Stereo Output Port on the SPEEDY-33.
- The on-board microphones will be used as the input to the DSP board.

Once the setup of the components is finished, open LabVIEW 8.5. You should get a window as shown in Figure 1.2.

The next step is to choose a target for the DSP program to be downloaded to. Using Lab-VIEW 8.5, it is very simple to do so by choosing *DSP Project* from the *Targets* pull-down menu and clicking on the *Go* button to open the DSP Project Wizard.

Select the *New DSP project, blank VI* from the *Project type* pull-down menu on the *Define project information* page to create a new project with an empty VI as shown in Figure 1.3 and then click *Next*.

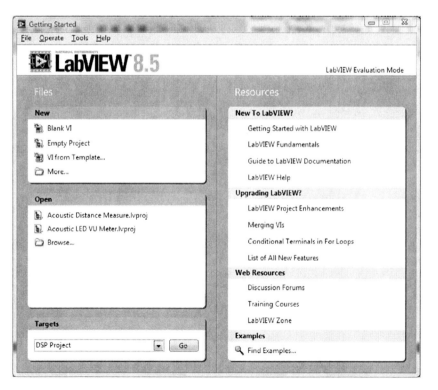

FIGURE 1.2: LabVIEW 8.5 Getting Started window.

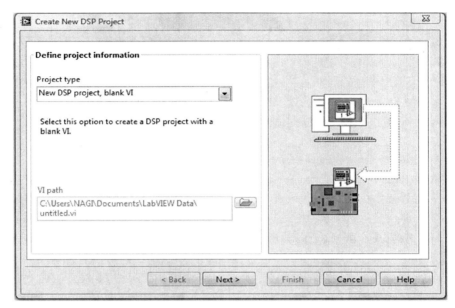

FIGURE 1.3: Creating a new DSP project.

Because the SPEEDY-33 is used here, select SPEEDY-33 from the *Target type* pull-down menu on the *Select target type and input/output resources* page as shown in Figure 1.4.

Click the *Finish* button to close the wizard and create the DSP project with a blank VI.

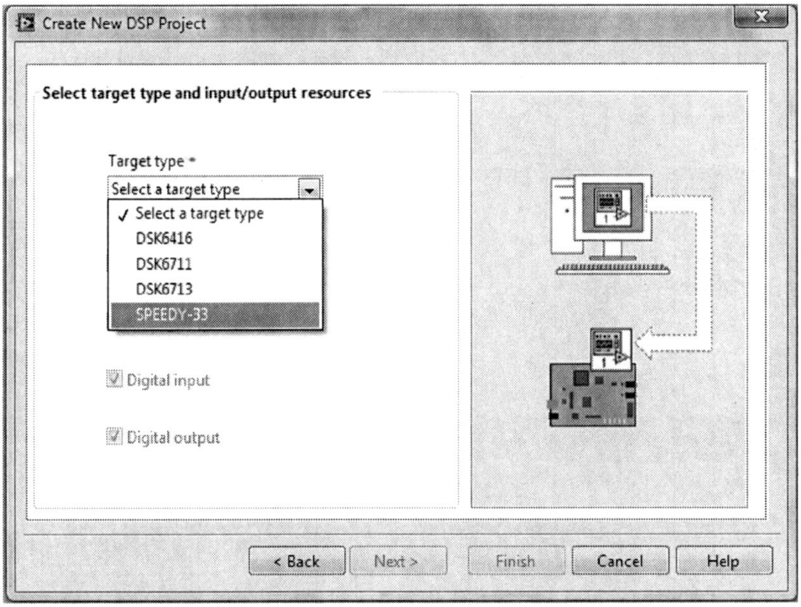

FIGURE 1.4: Selecting SPEEDY-33 Target.

As a result, two blank windows will appear which form the basic elements of a LabVIEW application also called Virtual Instrument (VI).

The gray color window is called the Front Panel. It is the place where the Graphical User Interface (GUI) is built.

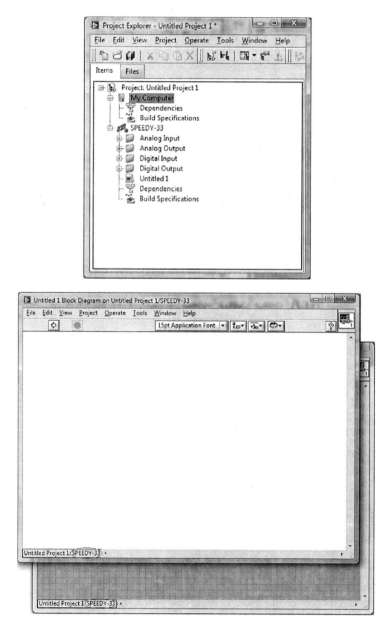

FIGURE 1.5: Front Panel (Dark window) and Block Diagram (White window) in LabVIEW.

The white color window is called the Block Diagram. It is where the functionality of the program is defined.

You can toggle between the two windows by pressing Ctrl + E.

The LabVIEW DSP module also shows the target that has been selected in the left bottom of the window.

1.7   A/D AND D/A CONVERSION

In this section, you will develop a simple VI. As a starting point, you will read an analog input from the A/D channel and play it back using the D/A channel.

To insert objects in the Block Diagram window, right-click anywhere on the white color window. By doing that, the Functions Palette will appear as shown in Figure 1.6. It contains all functions (VIs) that you may need to insert in the Block Diagram to develop an application. You can also click on the thumbtack 📌 (upper left corner) to tack down the palette.

To acquire an analog input from the A/D channel, you should click on the Elemental I/O subpalette ▦ and bring out the Analog Input Elemental I/O node 📇 and place it on the block diagram.

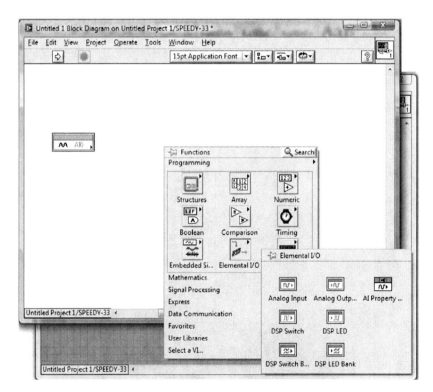

FIGURE 1.6: Using Function Palette to insert objects in the Block Diagram window.

By right-clicking on the Analog Input node and choosing *Properties*, the configure Elemental I/O panel opens up which allows you to configure the analog input.

By clicking on the configuration tab, you can select from a variety of sampling rates (8,000->48,000 Hz). Choose a sampling rate of 48 kHz and a framesize of 256 as shown in Figure 1.7.

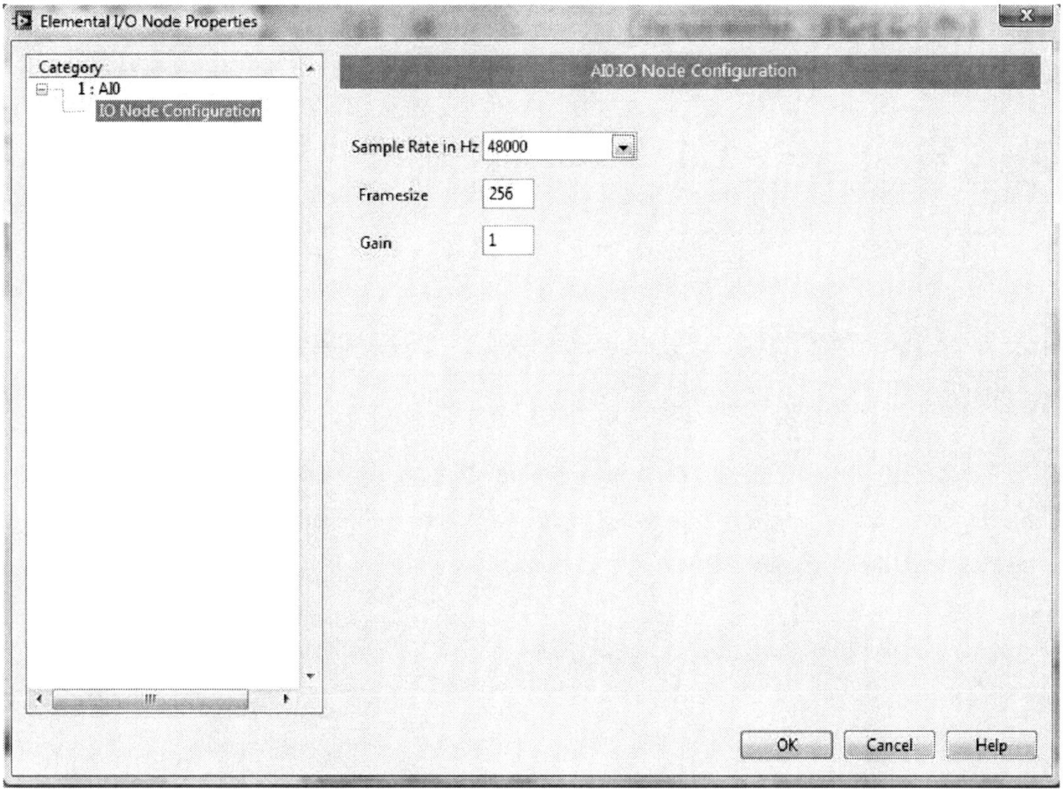

FIGURE 1.7: Configuring A/D Input node.

Click on the OK button to close the configuration panel for Analog Input and return to the block diagram.

The Analog Input block is equivalent to an A/D converter. The Analog Output block is equivalent to the D/A converter.

Similarly, repeat the same steps to place and configure an Analog Output Elemental I/O node on the block diagram.

Finally, connect the output terminal of Analog Input to the input terminal of the Analog Output. To do so, move the mouse over the left (right) channel output of the Analog Input block; the mouse curser will change to a wire spindle. Left-click and drag the wire to the left (right) channel input of the Analog Output block as illustrated in Figure 1.8.

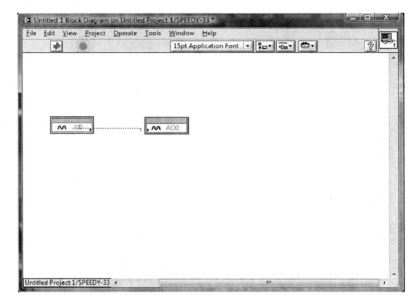

FIGURE 1.8: Connecting Analog Input to Analog Output.

Because the program should keep reading input data continuously without stopping, a While Loop should be added to the VI by going to the Structures subpalette and selecting While Loop as shown in Figure 1.9.

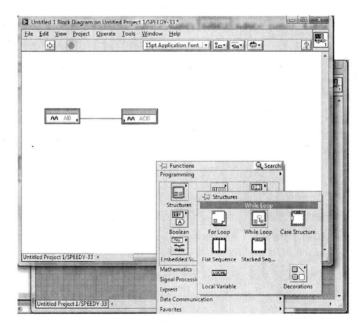

FIGURE 1.9: Selecting the Structures Subpalette in Block Diagram window.

Place the While Loop around the two blocks by left-clicking on the top left corner and dragging it to the bottom right. A Stop button should appear wired to the conditional terminal of the loop. It will also appear on the front panel. This button is used to abort execution of the VI while it is running. The final VI should look similar to the one shown in Figure 1.10.

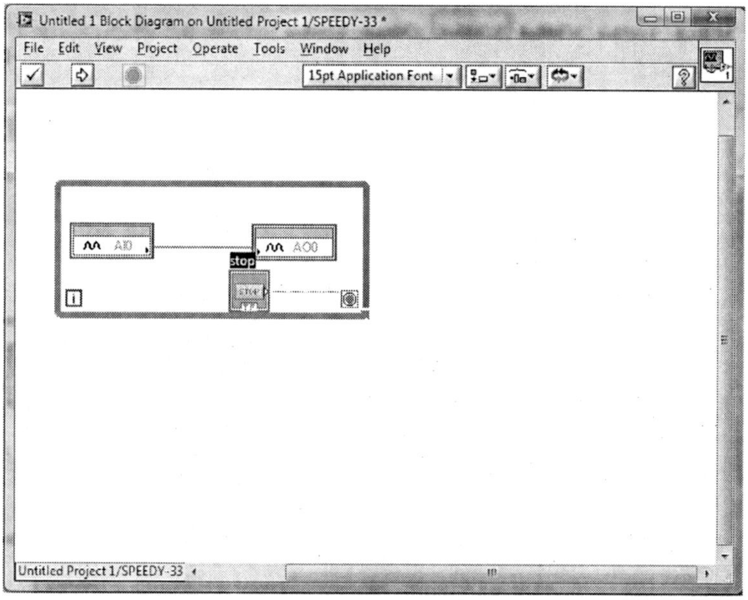

FIGURE 1.10: VI with While Loop in Block Diagram window.

To run the VI, click on the Run ⇨ button in the upper left corner of the Block Diagram window. The LabVIEW DSP Module Status Monitor window opens as shown in Figure 1.11. It should show that the program is compiled with 0 errors and that it is downloading to the target.

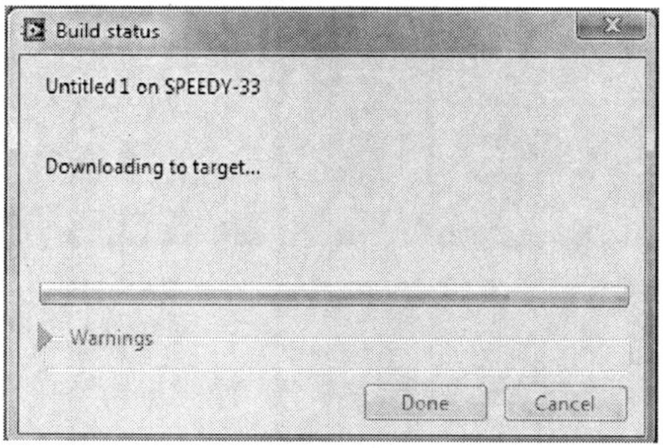

FIGURE 1.11: LabVIEW DSP status monitor window.

1. Try speaking through the onboard microphones while listening through the headphones. Can you pick up the sound?
2. Try to get further form the board. Can you still pick up the sound?

Stop running the program by clicking the Stop button in the Front Panel window as shown in Figure 1.12.

FIGURE 1.12: Front Panel window with Stop button.

To make the microphone more sensitive, you can add a gain to the Analog Input before feeding it to the Analog Output.

3. Try this by modifying the VI to amplify the input signal by 4. Is the sound clearer?

Hint: Use a function in the Numeric Subpalette

1.8   DISPLAYING SIGNALS IN TIME DOMAIN

To be able to see the shape of the input signal in time, one will have to use waveform graphs.

To set up a basic time-domain display using a Waveform Chart, go to the Front Panel and right-click on it to bring up the Controls palette. From the Graph subpalette, choose Waveform Graph.

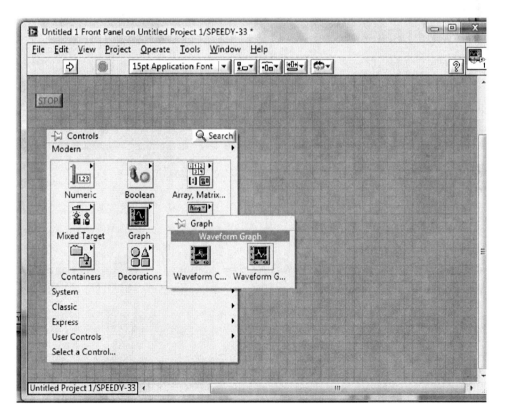

FIGURE 1.13: Controls Palette and Graph Subpalette in Front Panel window.

Place the Waveform Graph on the Front Panel. Double-click on the Waveform and change the label to Input Signal as shown in Figure 1.14.

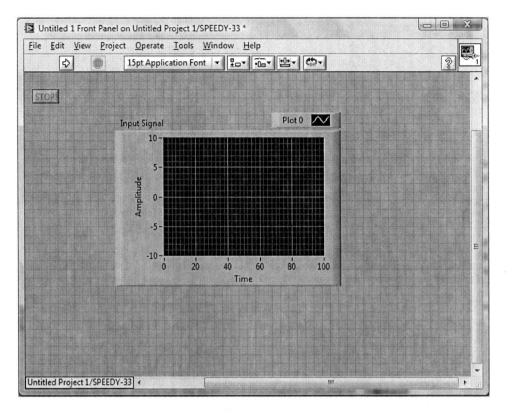

FIGURE 1.14: Waveform Graph for the left and right input signals.

If you go to the Block Diagram window, you should notice that there should be an icon corresponding to the Waveform Graph inserted in the Front Panel window. Wire the Output of the Analog Input to Input Signal graph.

Run the VI. Run the program and speak into the microphones. You should be able to see your voice changing with time in the waveform graph.

4. Try to whistle into the onboard microphone. What do you notice in the graph? Comment.

Hint: Whistling gives a single tone.

1.9   DEBUGGING TECHNIQUES

Debugging is an efficient tool in program, as it allows you to watch the data flow of a program, find errors, and correct them.

There are different debugging techniques:

Finding errors. To list errors, click on any broken arrow.

5. Test this feature by making an intentional error in your program and click on the broken arrow to read the error. Comment.

Execution highlighting. This animates the diagram and traces the flow of the data, allowing you to view intermediate values *(available only in LabVIEW 8.5 and not LabVIEW DSP Project).*

6. Test this feature by running your program and clicking on the light bulb on the toolbar. Comment. making an intentional error in your program and click on the broken arrow to read the error. Comment.

Probe. This is used to view values in array, which is a list of input values and clusters.

7. Test this feature by running your program and right-clicking on any wire to set a probe. Making an intentional error in your program and click on the broken arrow to read the error. Comment.

Breakpoint. This sets pauses at different locations on the diagram.

8. Test this feature by clicking on wires or objects with the Breakpoint tool to set breakpoints. Make an intentional error in your program and click on the broken arrow to read the error. Comment.

TO PROBE FURTHER

- *LabVIEW demonstrations in Connexions:* http://cnx.org/content/m11821/latest/
- *LabVIEW Graphical Programming Course:* http://cnx.org/content/col10241/latest/

REFERENCES

1. Nasser Kehtarnavaz and Namjin Kim, *Digital Signal Processing System-Level Design Using Lab-VIEW*, Elsevier/Newnes, Amsterdam/Boston, 2005.
2. Signal Processing Engineering Educational Device for Youth, *NI SPEEDY-33 User Manual*, National Instruments, Austin, TX.

•　　•　　•　　•

CHAPTER 2

Applications Using LEDs and Switches Using the SPEEDY-33

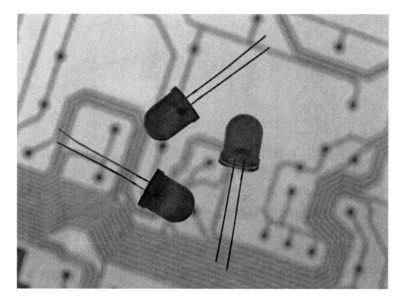

2.1 OVERVIEW

In this chapter, you will get familiar with the digital input and output ports of the board. The inputs are the switches, and the outputs are the LEDs. In particular, you will be implementing an LED VU (VU stands for Volume Units) meter that shows the relative strength of an acoustic signal. Another application will be to simulate a roulette game.

2.2 BACKGROUND
2.2.1 LEDs and Switches

Though the primary focus of DSP applications is streams of data such as audio and speech signals, there are applications that benefit from digital inputs and outputs. Concepts such as triggering and interrupts can be illustrated with the use of digital inputs, while the LEDs can be used to build many things ranging from a simple error indicator to an LED VU meter.

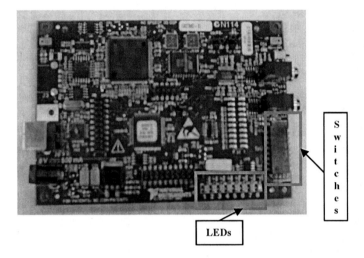

FIGURE 2.1: Speedy-33 board.

As seen in Figure 2.1, the NI SPEEDY-33 comes well equipped with digital I/O. It has eight onboard LEDs for digital output and eight DIP (Dual In-line Package) switches for digital input. The digital input and output lines of the NI SPEEDY-33 can be individually configured from the LabVIEW DSP Module. The LEDs outputs are enabled by writing a 1 to the appropriate bit of the LED port. The switch inputs are accessed by the DSP through reading the appropriate bit of the switch input port. In addition, the LED output state and the switch inputs are connected in parallel to pins on the simple expansion digital I/O connector, which could be useful for robotic applications.

2.3   EXPERIMENT

Now that you have an idea about the digital input/output ports of the Speedy-33 board, you are ready to begin the experiment and implement some fun applications using LEDs and switches.

2.3.1   Application 1: Controlling LEDs Using Switches

As a starting application, you will implement a VI to check the status of the DIP switches and accordingly manipulate the corresponding LEDs; i.e., if DIP switch 1 is ON, turn LED1 ON, and so on.

This is done by choosing from the *Functions* palette the *DSP Switch* node found in *Elemental I/O*, as shown in Figure 2.2.

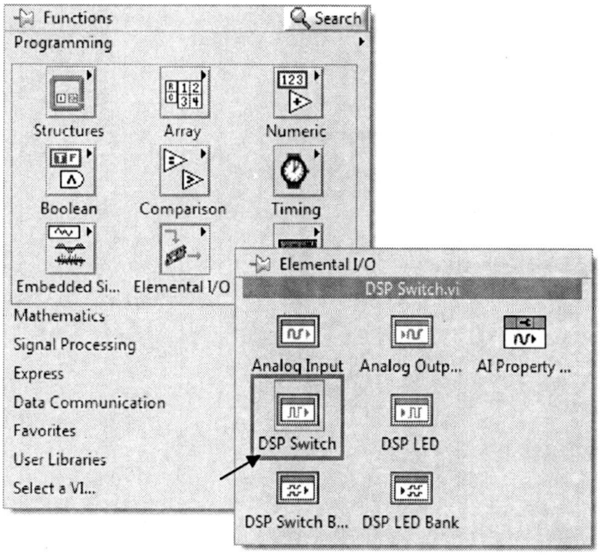

FIGURE 2.2: DSP switch node selection.

By placing the *DSP Switch* node in the window, you can read the status of a selected DIP switch.

Connect the output of the Digital Input node to an indicator, by right-clicking on the output of the Digital Input node and choosing Create → Indicator as shown in Figure 2.3.

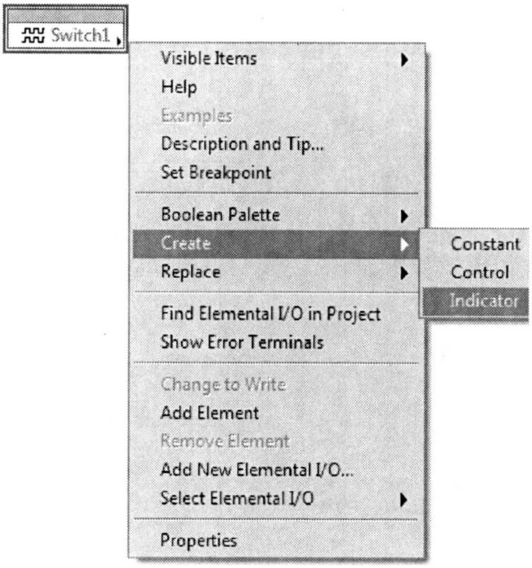

FIGURE 2.3: Selecting indicator for DSP switch.

The Front Panel and Block Diagram windows should show a Boolean indicator as shown in Figures 2.4 and 2.5.

FIGURE 2.4: Resulting block diagram.

FIGURE 2.5: Resulting front panel.

Place a While Loop so that the VI keeps running when you press the Run button.

1. Try pressing different DIP switches, while watching the indicator in the Front Panel. What happens? Comment.

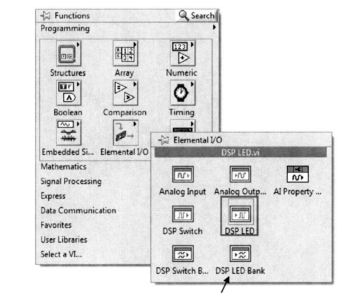

FIGURE 2.6: Selecting DSP LED.

You can configure the LEDs in a similar way:

This is done by choosing from the *Functions* palette the *DSP LED* node found in *Elemental jI/O*, as shown in Figure 2.6.

By placing the *DSP LED* node in the window, you can control the status of any LED.

Connect the input of the DSP LED node to a control, by right-clicking on the input of the DSP LED node and choosing Create → Control, as shown in Figure 2.7.

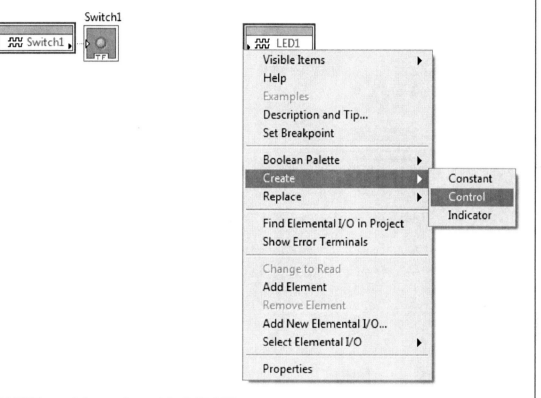

FIGURE 2.7: Selecting Control for DSP LED.

The Front Panel and Block Diagram windows should show a Boolean control as shown in Figures 2.8 and 2.9.

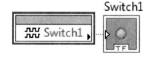 

FIGURE 2.8: Final resulting block diagram.

FIGURE 2.9: Final resulting Front Panel.

2. Try pressing the LED control in the Front Panel, while watching the LEDs on the board. What happens? Comment.

Remove the Control from the DSP LED input and connect it to the output of the DSP Switch as shown in Figure 2.10.

FIGURE 2.10: Digital Input connected to Digital Output.

3. Try pressing the DIP switches on the board, while watching the LEDs on the board. What happens? Comment.

Modify the VI so that you check the status of the eight DIP switches and accordingly manipulate the eight corresponding LEDs. Your final VI should be similar to the figure shown in Figure 2.11.

FIGURE 2.11: Eight LEDs controlled by eight DIP switches.

Run the VI and press different switches and notice the corresponding LEDs that turn on. Comment.

2.3.2   Application 2: Audio LED VU Meter

For audio systems or recorders, an audio VU (VU stands for Volume Units) meter is a device that indicates the relative levels of the audio being recorded or played.

To implement an Audio LED VU meter using the Speedy-33 board, you should acquire an audio signal and compare its energy level with different levels. If the energy is higher than a certain level, a corresponding LED will be turned on.

The first step to do is to acquire the audio signal by placing an *Analog Input* node in the Block Diagram window, with a sampling rate of 8 kHz.

To get the energy level of the signal, look for the RMS block ⊞, which you can find in the *Signal Processing* → *Time Domain* sub-palette as shown in Figure 2.12.

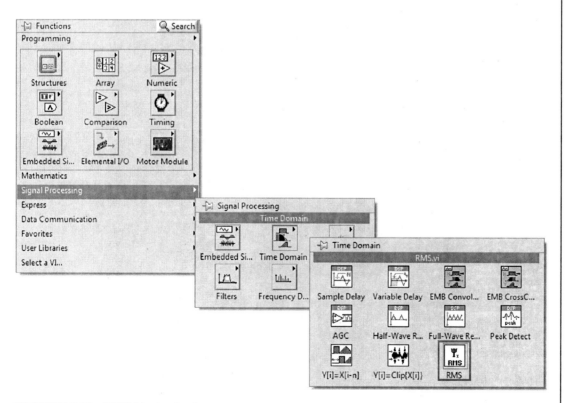

FIGURE 2.12: RMS block selection.

Connect the output of the analog input block to the input of the *RMS* block. Connect the output of the *RMS* block to an indicator, so that you can see the energy level of the incoming signal, as shown in Figure 2.13.

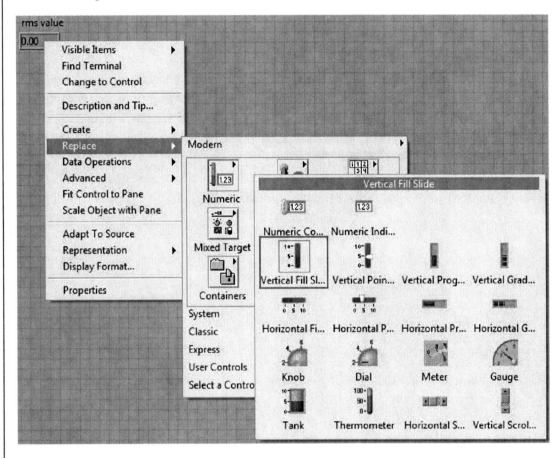

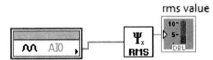

FIGURE 2.13: Analog Input, RMS, and indicator connections.

The RMS block computes the root mean square (rms) of the input sequence x using the following equation:

$$\psi_x = \sqrt{\frac{1}{n}\sum_{i=0}^{n-1} x_i^2}$$

Next, you should compare the value of the energy level with different constants. If the energy level is higher than a constant, then the corresponding LED should be turned on.

To do so, find the *Greater* block ▷ found in the *Comparison* sub-palette ▷ as shown in Figure 2.14.

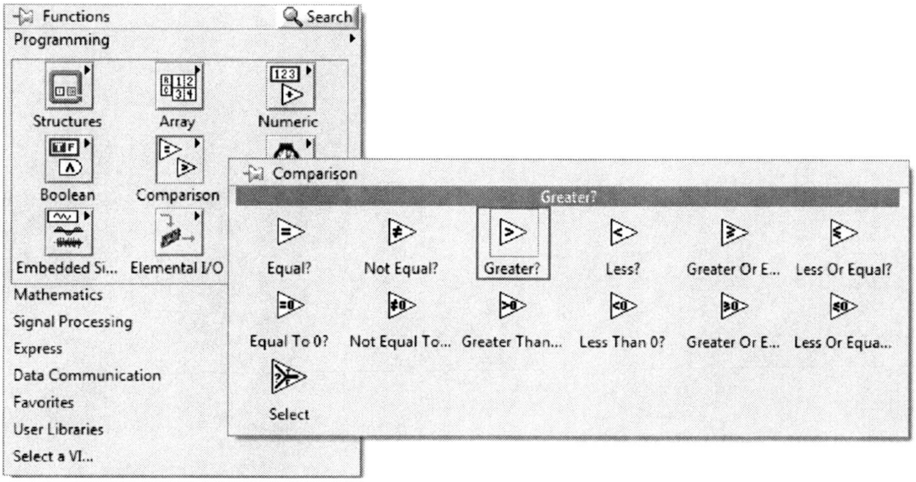

FIGURE 2.14: Greater Block selection.

Place eight *Greater* blocks, each block corresponding to an LED.

Connect the output of the *RMS* block to the input *x* of the *Greater* blocks. Connect a constant to the second input *y* of each of the Greater blocks, by right-clicking on the input *y* and choosing *Create* → *Constant* as shown in Figure 2.15.

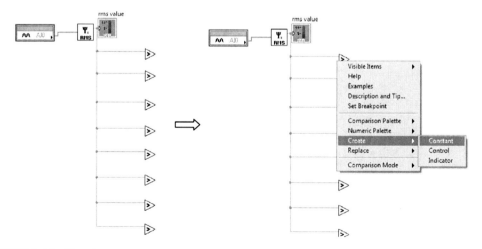

FIGURE 2.15: Creating a constant.

Place eight constants for each *Greater* block. Choose the constants to be 0.3 0.4, 0.5, 0.6, 0.7, 0.8, 0.9, and 1 as shown in Figure 2.16.

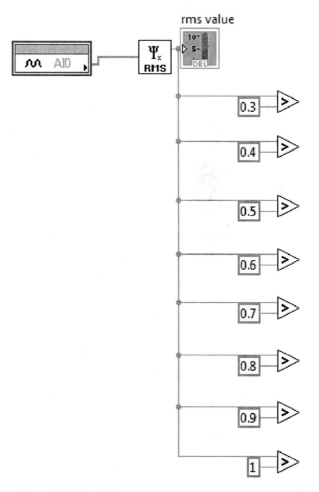

FIGURE 2.16: Comparing with eight different constants.

The last step is to compare the signal energy level of the incoming audio signal with a certain level and turn on an LED depending on the comparison result.

To do so, connect eight *DSP LEDs*, corresponding to eight different LEDS, to the output of the Greater blocks. Place a While Loop.

The final VI should be similar to the one shown in Figure 2.17.

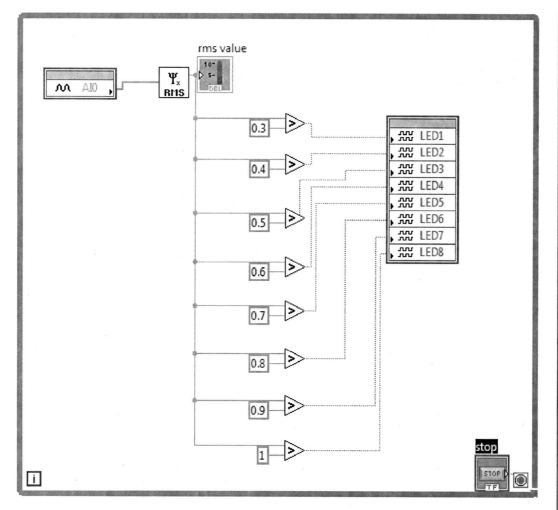

FIGURE 2.17: Audio level meter VI.

4. Run the VI and comment on the status of the LEDs when you speak.

5. Try to vary the sensitivity of the VU meter by placing a Divider function and connecting its two inputs to the RMS node output and to a control. The output of the multiplier will be connected to the input of the Greater functions. The control range should be between 0 and 10,000.

CHAPTER 3

Noise Removal

3.1 OVERVIEW

In this chapter, you will get familiar with designing and implementing digital filters using the Lab-VIEW DSP Module and the SPEEDY-33 DSP board. As a starting point, you will get familiar with the frequency response of filters. Later, you will design your own digital filter to remove noise from a corrupted signal.

3.2   BACKGROUND

As mentioned in Chapter 1, the most common representation of signals and waveforms is in the time domain. However, most signal analysis techniques work only in the frequency domain. When it is first introduced, the concept of the frequency domain representation of a signal is somehow difficult to understand. This chapter attempts to explain the frequency domain representation of signals. The frequency domain is simply another way of representing a signal.

3.2.1   Sinusoidal Signals and Frequency

A *sinusoid* is any function of time having the following form:

$$x(t) = A \sin (2\pi f t + \phi)$$

where A is the peak amplitude (nonnegative), t is time (sec), f is frequency, and ϕ is initial phase (radians).

An example of a sinusoidal signal (sinusoid) is shown below:

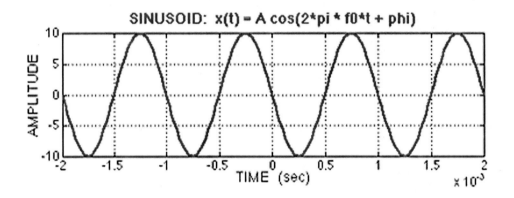

The term "peak amplitude" is often shortened to "amplitude."

The "phase" of a sinusoid normally means the "initial phase." Another term for initial phase is *phase offset*.

Note that Hz is an abbreviation for hertz, which physically means cycles per second.

Although we only examined a sinusoidal waveform, it is relevant to all waveforms because any non-sinusoidal waveforms can be expressed as the sum of various sinusoidal components.

The following two sinusoids have frequencies of 300 and 500 Hz, respectively:

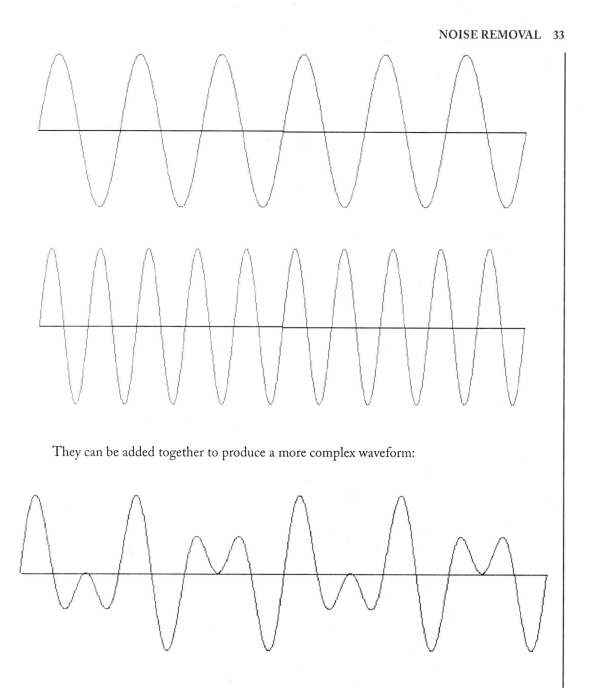

They can be added together to produce a more complex waveform:

This is very important because, in general, one can represent any complex continuous wave as a combination of simple sine waves with different frequencies. In other words, any signal (voice, music, etc.) is composed of different frequencies. For example, the voice signal is composed of frequencies ranging from 300 to 3,300 Hz, while music frequencies range from 20 Hz to 20 kHz. One can take advantage of this fact to eliminate unwanted frequencies, as it will be shown in this experiment.

The frequency-domain display shows how much of the signal's energy is present as a function of frequency; in other words, for each frequency value, the frequency-domain representation displays the amplitude (amplitude spectrum) and phase (phase spectrum) of the sinusoidal component present in the signal at that frequency. For a simple signal such as a sine wave, the frequency domain representation does not usually show us much additional information. However, with more complex signals, the frequency domain gives a more useful view of the signal.

To summarize, any complex periodic waveform can be decomposed into a set of sinusoids with different amplitudes, frequencies, and phases. The process of doing this is called Fourier analysis, and the result is a set of amplitudes, phases, and frequencies for each of the sinusoids that make up the complex waveform. Adding these sinusoids together again will reproduce exactly the original waveform. A plot of the amplitude of a sinusoid against frequency is called an amplitude spectrum, and a plot of the phase against frequency is called phase spectrum.

3.2.2   What Is a Filter?

A filter is a device that accepts an input signal and passes or amplifies selected frequencies while it blocks or attenuates unwanted ones.

For example, a typical phone line acts as a filter that limits frequencies to a range considerably smaller than the range of frequencies human beings can hear. That is why listening to CD-quality music over the phone is not as pleasing to the ear as listening to it directly using a CD player.

Filters can be *analog* or *digital*.

A *digital filter*, which will be covered in this chapter, takes a digital input, gives a digital output, and consists of digital components. In a typical digital filtering application, software running on a digital signal processor (DSP) reads input samples from an A/D converter, performs discrete mathematical manipulations for the required filter type to possibly eliminate some frequencies, and outputs the result via a D/A converter.

An *analog filter*, by contrast, operates directly on the analog inputs and is built entirely with analog components, such as resistors, capacitors, and inductors.

The frequency response of a filter is the measure of the filter's response (filter output) to a sinusoidal signal of varying frequency and unit amplitude at its input.

There are many filter types, but the most common ones are *lowpass*, *highpass*, *bandpass*, and *bandstop*. They are shown below along with their frequency response.

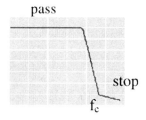

A *lowpass* filter allows only low frequency signals (below some specified cutoff frequency, f_c) through to its output, so it can be used to eliminate high frequencies.

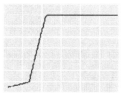

A *highpass* filter does just the opposite, by rejecting only frequency components below some cutoff frequency, f_c. An example highpass application is cutting out the audible 60 Hz AC power "hum", which can be picked up as noise accompanying almost any signal in the United States.

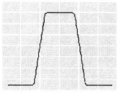

The designer of a cell phone or any other sort of wireless transmitter would typically place an analog *bandpass* filter in its output RF stage, to ensure that only output signals within its narrow, government-authorized range of the frequency spectrum are transmitted.

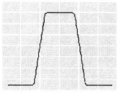

Engineers can use *bandstop* filters, which pass both low and high frequencies, to block a pre-defined range of frequencies in the middle.

3.3   EXPERIMENT

Now that you have an idea about frequency and filters, you are ready to begin the experiment.

3.3.1   Part A. Generating Signals Using Sinusoids

Using LabVIEW, you can easily generate sinusoids signals.

Once you open LabVIEW DSP Module, create a blank VI. In the Block Diagram window, place an EMB Sine Waveform  Express VI located on the *Functions >> Embedded Signal Generation* palette as shown in Figure 3.1.

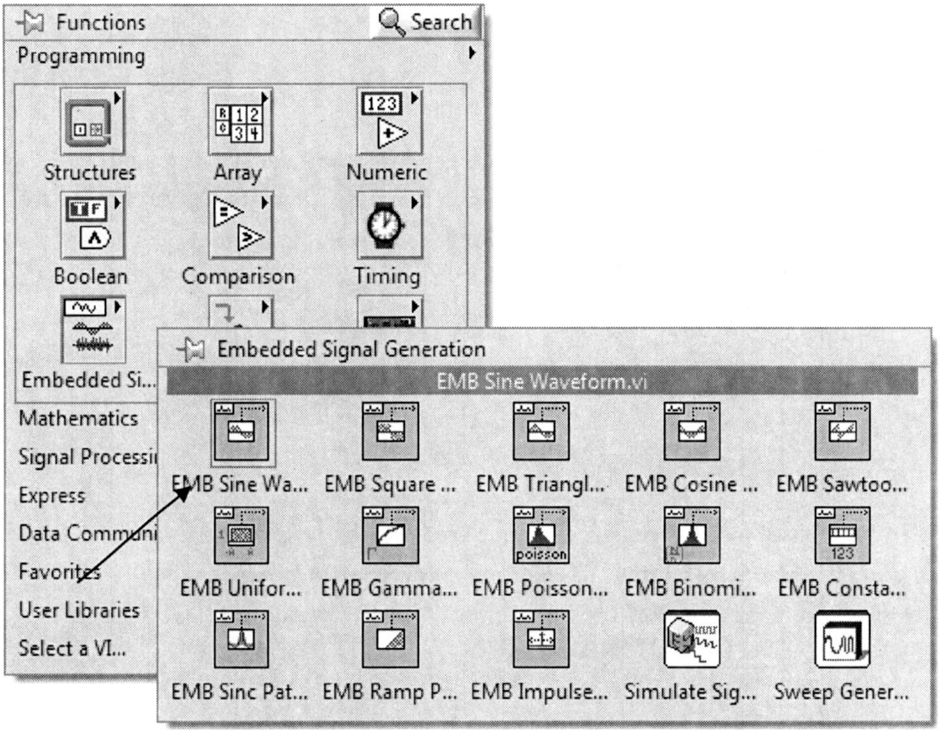

FIGURE 3.1: Choosing EMB Sine Waveform.

Connect two control knobs to the amplitude and frequency nodes of the EMB Sine Waveform block.

Add a Graph to show the sine wave.

1. Observe the sinewave on the graph while varying the Amplitude and Frequency knobs.

A new periodic signal can be generated by combining together sinusoids having harmonically related frequencies.

For example, a square wave $x(t)$ can be generated from sinusoids using:

$$x(t) = \frac{8}{\pi} \cos(2\pi f_0 t - \pi/2) + \frac{8}{3\pi} \cos(2\pi 3 f_0 t - \pi/2)$$

2. Using the above information, generate a square wave using EMB Sine Waveform.

3.3.2   Part B. Plot Signals in Frequency Domain

To display the frequency content of an audio signal, construct first the block diagram shown below. Set the sampling frequency in the input and output to 8,000 Hz.

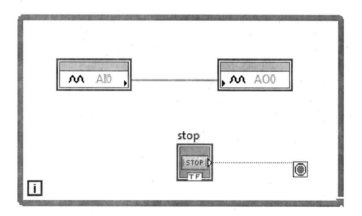

FIGURE 3.2: Block diagram.

Add a graph corresponding to the audio input.

Change the xlabel name of the graph in the Front Panel window to frequency instead of time. Right-click on both graphs and choose *Properties*. Because the framesize that was chosen is 256,

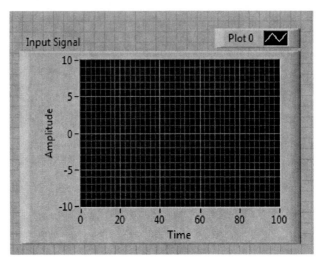

FIGURE 3.3: Graph for Input Signal.

under the *Scales* tab, change the range of the frequency axis to 0–255. The last frame corresponds actually to the sampling frequency, which is 8,000 Hz in our case. Therefore, in the Multiplier field, insert 31.3725 which is 8,000/256 as shown below.

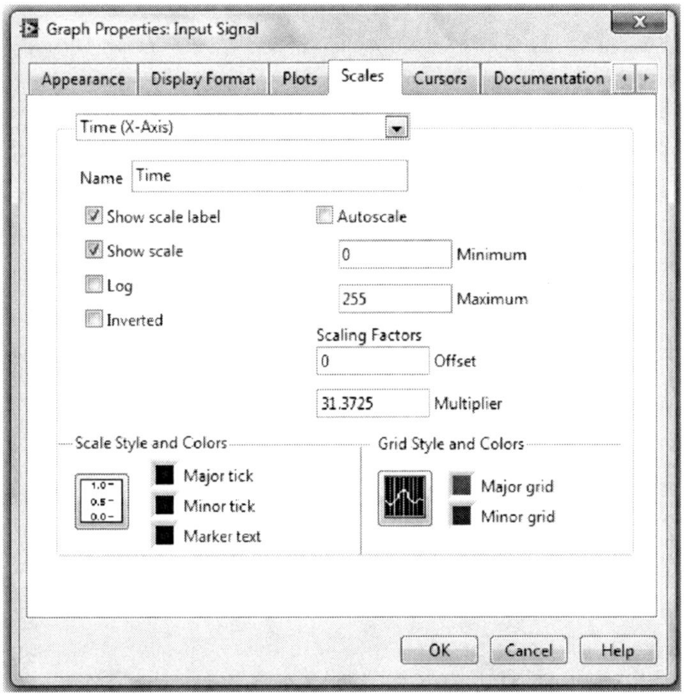

FIGURE 3.4: Scaling the xlabel of the graph.

3. To display the frequency content of the audio signal use the Fast Fourier Transform (FFT), go to the block diagram window, and connect the Spectral representation of the input signal to the graph. Look in the Signal Processing subpalette for the FFT block.

4. Run the VI and notice the frequency range of your voice.
5. Try to whistle and notice the corresponding frequency spectrum. Comment.
Change the pitch (frequency) of the whistle and examine how it affects the frequency spectrum. Try experimenting with different notes.

3.3.3   Part C. Frequency Response

In this part, you will design a lowpass digital filter and then you will draw its frequency response, in other words you will draw the output amplitude at different frequencies of the input signal.

Add an EMB Sine Waveform.

Insert a filter located on the *Functions >> Signal Processing >> Filters >> Filter* palette as shown in Figure 3.5.

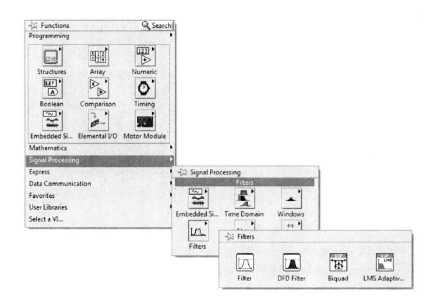

FIGURE 3.5: Choosing filter in LabVIEW.

Double-click on the filter VI and configure the filter to have a lowpass filter type with a cutoff frequency of 500 Hz.

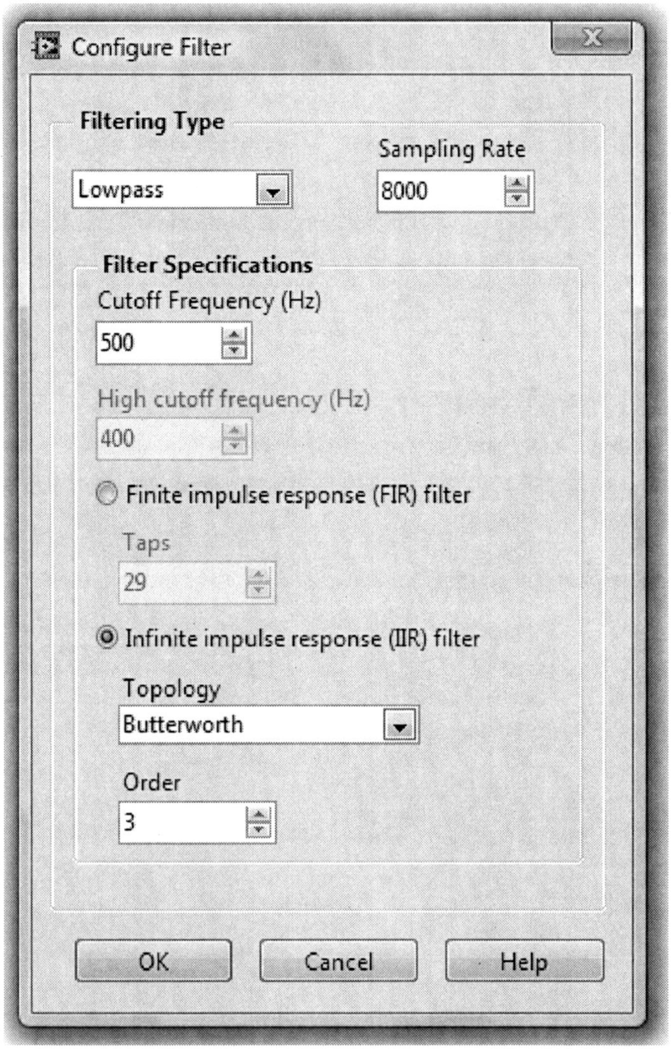

FIGURE 3.6: Configuring filter.

In the Front Panel, place two *knob controls* and a *Waveform Graph* as shown below. Name the two knobs *Amplitude* and *Frequency*. Make sure to let the Frequency knob vary from 0 to 1,000.

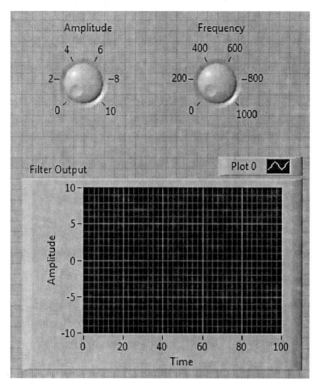

FIGURE 3.7: Front Panel view.

Go back to the Block Diagram and make all the necessary wiring. At the end, you should obtain the following:

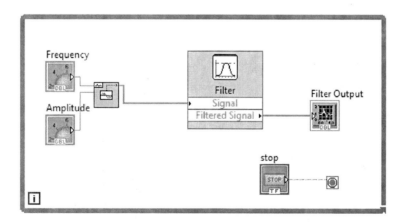

FIGURE 3.8: Block Diagram connections.

6. Set the amplitude of the input sinewave to 6. Vary the frequency of the input sinewave from 10 to 1,000 Hz. Observe the output amplitude at each frequency and plot the resulting frequency response.

7. Repeat the previous steps for a highpass filter with the same cutoff frequency.

The cutoff frequency is the frequency at which the power of the signal is reduced by half. In analog filter design, the cutoff frequency is usually defined to be where the amplitude is reduced to 0.707.

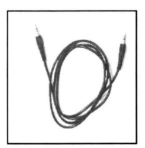

8. Find the frequency at which the output amplitude is 0.707*input amplitude. Does it correspond to 500 Hz? Explain.

3.3.4   Part D. Noise Removal

In this part, you will have to design your own system to remove noise from a corrupted signal.

To do so, download the corrupted audio file from the blackboard and play it. The file was corrupted by adding a sinusoidal tone to it. You should be able to hear a voice corrupted with the sinusoidal noise. Connect the output sound card of the PC to the input of the DSP board using the double jack stereo cable. The corrupted audio file is now the input signal for the SPEEDY-33 board.

9. Find the frequency of the noise by looking at the FFT of the input signal.

Hint: In the frequency spectrum, find the peaks with constant amplitudes.

10. Design a bandstop filter to eliminate the noise components and observe the frequency spectrum of the filtered signal.

Hint: A lowpass and a highpass filter can be combined to produce a bandstop filter.

CHAPTER 4

Music Equalizer

4.1   OVERVIEW

In this chapter, you will design and implement a five-band music equalizer. Each band is controlled by different filters. Each filter controls the magnitude of a certain frequency range by adjusting the gain of the filter.

4.2   BACKGROUND

4.2.1   Music Equalizer

Frequency is the number of vibrations or oscillations per second, which is measured in cycles per seconds or hertz. It is said that the human ear can perceive a range of sound frequencies from 20 Hz (20 cycles per second) to 20 kHz (20,000 cycles per second), but of course, that does not apply to all of us. Audio equalizers allow you to adjust levels at set of points in a range of frequencies.

Each set of point is called a band, and the more bands available for adjustment within a frequency range, the more precise the audible adjustments become. Many digital audio playing programs include an equalizer with anything from two bands, such as treble and bass sliders, up to over 100 bands each adjusting a certain frequency and surrounding frequencies.

Standard bass and treble tone controls are broadband devices that have the greatest effect at the frequency extremes—that is, the highest highs and the lowest lows. While this is fine for touching up the response in your car stereo, it is not very precise, as it cannot control small range of frequencies. A better way is to divide the audible spectrum into five or more frequency bands and allow adjustments to each band via its own boost/cut control. Instead of broad adjustments of treble, bass, and maybe the midrange, one has independent control over the low bass, mid-bass, high bass, low midrange, and so forth.

The gain of a frequency can be boosted (an increase in gain of a frequency) or cut (reduction in gain of a frequency) by decibels (dB), increasing or decreasing the volume, noise, or intensity of that frequency. In this way, it is possible to target certain frequencies in a range to improve (or distort) the sound. Bass boost buttons or dials on hi-fi systems increase or "boost" in decibels the frequencies at the low end of the audible range, while treble adjustments increase in decibels the frequencies at the high end. Bass, treble, and mid-range frequency adjustments differ depending on the type of music you are listening to. To create the best sound for techno and dance tunes, higher bass and treble frequencies are advisable, but the same frequency settings would not usually suit listening to classical or metal. This is because different musical instruments (including the human voice) create frequencies located in specific areas within the audible range.

4.3   EXPERIMENT

4.3.1   Music Equalizer

After starting LabVIEW Embedded DSP Module and changing the target to the SPEEDY-33 board, create a New Blank VI.

Open the Block Diagram and place the Elemental I/O ≫ Analog Input node in it.

Double-click the Analog Input node and change the Resource under the General Tab to be one-channel multiple samples and check that the Sample Rate is set to 8,000 Hz and Framesize is 512. One channel means that the output is mono, and multiple samples means that more than one sample is taken each sampling time and is stored in a buffer size of 512 in our case. Repeat the same for the Analog Output node.

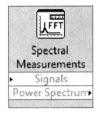

FIGURE 4.1: FFT power spectrum.

Place a Spectral Measurements Express VI and double-click on it to select Power Spectrum from the type of Measurement.

Rearrange the objects in the Block Diagram so that you have plenty of space between them.

Next, the incoming signal should be partitioned into different bands, so that one can control different frequency regions. The more regions available for adjustment within a frequency range, the more precise the audible adjustments become. In this experiment, six regions will be used: 0–50, 50–100, 100–250, 250–750, 700–1,500, >1,500 Hz.

To do that, in the block diagram, place the Signal Processing ≫ Filters ≫ Filter VI.

To choose the frequency range 0–50, double-click in the VI and configure a lowpass FIR filter with the following settings:

- Type: Lowpass
- Sampling frequency: 8,000 Hz.
- Cutoff frequency: 50 Hz
- FIR filter: Elliptic, order 7

The Elliptic type and Order 7 were chosen, as they would result in a sharp filter, but any other type and order could still be chosen.

Connect the output of the Analog Input to the input of the Filter as illustrated in Figure 4.2.

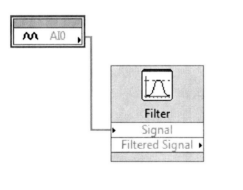

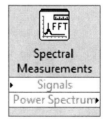

FIGURE 4.2: Connecting output of Analog Input to the input of the filter.

Repeat the same for the other bands, using a lowpass and a highpass filter in cascade with the lower frequency of the band being the cutoff of a highpass filter and the higher frequency of the region being the cutoff of a lowpass filter. For instance, for the region 50–100 Hz, use a highpass filter of cutoff 50 Hz and a lowpass of cutoff 100 Hz in cascade. It is easier to use a bandpass filter; however, LabVIEW DSP Module does not support with Filter VIs bandpass filters.

> 1. For the last region (>1,500 Hz), specify the properties of its corresponding filter and explain your choice.

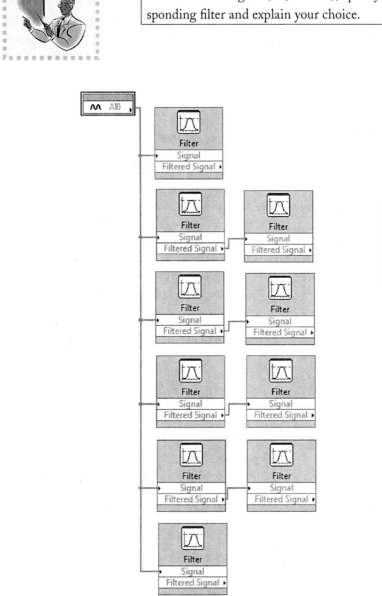

FIGURE 4.3: Connection of different filters in Block Diagram.

As shown in the Figure 4.3, the set of ten filters splits the analog signal into the six frequency bands shown on the front panel. The top filter is a lowpass filter that passes frequencies from 0 to 50 Hz. The four pairs of lowpass–highpass filters create bandpass filters of different ranges. The bottom filter is the filter you chose that passes frequencies above 1.5 kHz.

The next step is to control the output of each filter by either boosting the input signal to the filter or attenuating it.

To do that, go to the Front Panel and place six slide controls in the Front Panel. Place six slide indicators under each slide control. You should also add an additional knob control for the volume. Set the range of all slide controls to be between 0 and 2. The result should be similar to the one shown in Figure 4.4.

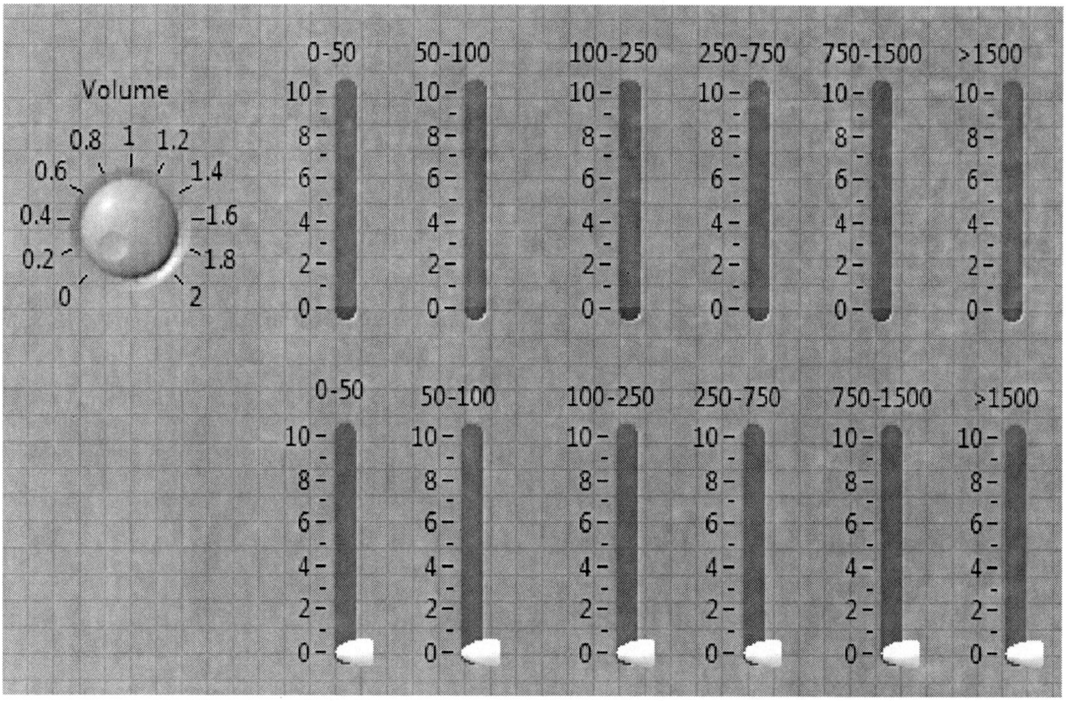

FIGURE 4.4: Music Equalizer Control Panel.

Arrange the Block Diagram as shown in Figure 4.5:

Put six Multiply functions and wire the output of each cascade filter with the Frequency Slide Controls into the Multiply Functions.

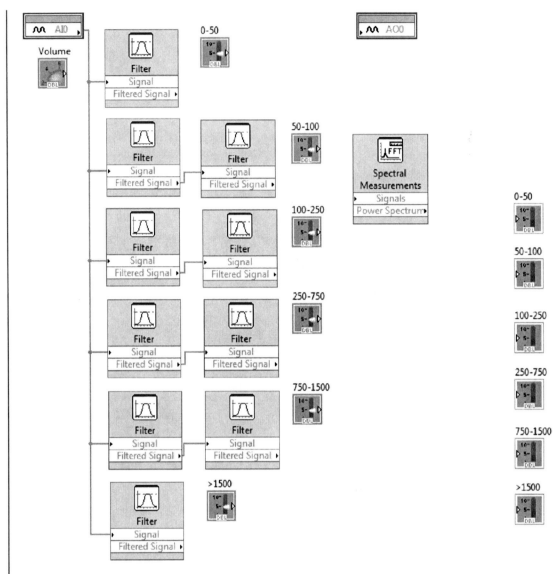

FIGURE 4.5: Arranging Block Diagram.

2. What are some of the possible uses of the Multiply Functions?

Add the six signals together to get the signal at all frequencies, by placing five Add functions, and wire the output of the last Add to the input of another Multiply Function. Connect the Volume Control to the second input of the Multiply Function as shown in Figure 4.6.

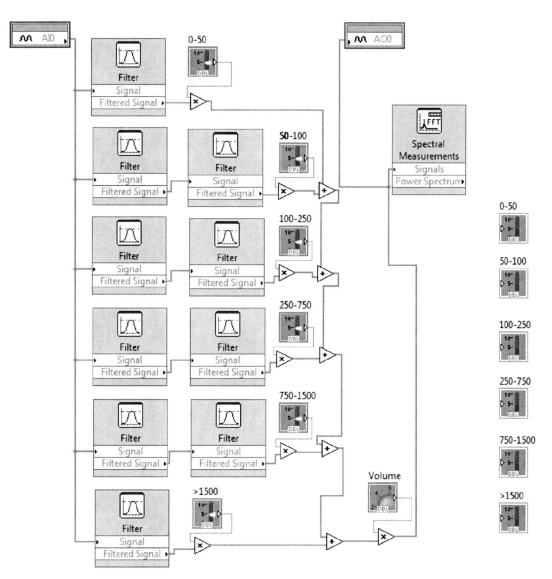

FIGURE 4.6: Adding signals together.

Next, the RMS value of the signal at a frequency within each frequency band should be displayed using the slide indicators on the front panel.

To do that, add an RMS block and place five index arrays from Arrays ≫ Index Array in the Block Diagram. Connect the output of the Spectral Measurements to the array of each of the five Index Arrays and the output of each array to the corresponding slide indicator, as shown in Figure 4.7:

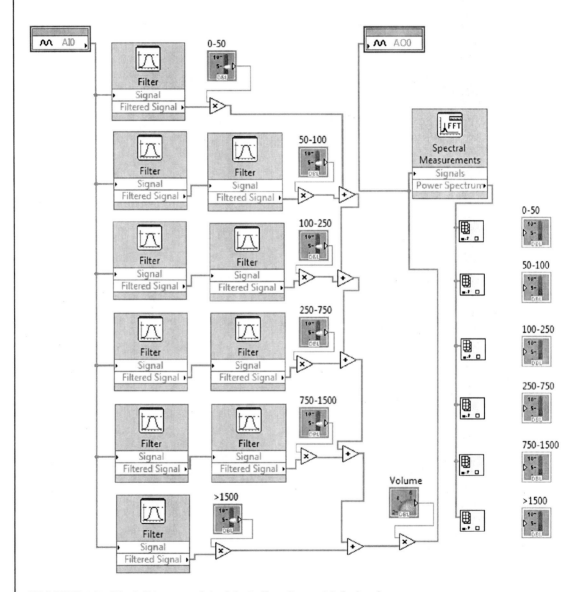

FIGURE 4.7: Block Diagram of the Music Equalizer with Index Arrays.

For each Index Array to take the RMS value of the signal at a frequency within each frequency band, the index of each Index Array should be 2, 4, 8, 32, 64, and 128.

3. Explain the reason behind the choices of constants for the index of each Index Array.

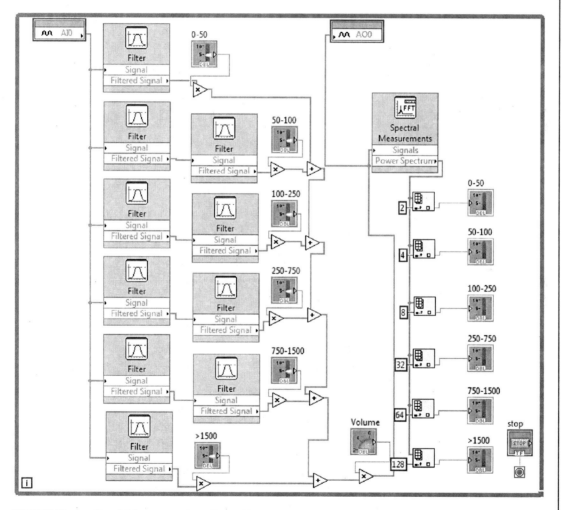

FIGURE 4.8: Final Music equalizer Block Diagram.

4. Start some music. Save and run the VI. You should first hear the music basically unfiltered, as all frequencies in the range 0–4,000 Hz are passed without attenuation. Start playing with the individual gains now to boost/attenuate the various frequency ranges by varying the gains. Comment.

CHAPTER 5

Telephone

5.1   OVERVIEW

In this chapter, you will generate a dual-tone multifrequency (DTMF) signal, which is a signal consisting of the sum of two pure sinusoids at valid frequencies. Touch tone telephones use DTMF signals.

5.2   BACKGROUND

5.2.1   How Does a Telephone Work?

Telephone touch pads generate DTMF signals to dial a telephone number. When any key is pressed, the tones of the corresponding column and row in Figure 5.1 are generated, hence dual tone. As an example, pressing the 5 button generates the tones 770 and 1,336 Hz summed together.

	1209Hz	1336Hz	1477Hz
697Hz	1	ABC 2	DEF 3
770Hz	GHI 4	JKL 5	MNO 6
852Hz	PRS 7	TUV 8	WXY 9
941Hz	*	OPER 0	#

FIGURE 5.1: Frequencies corresponding to rows and columns of phone.

The frequencies in Figure 1 were chosen to avoid harmonics. No frequency is a multiple of another, the difference between any two frequencies does not equal any of the frequencies, and the sum of any two frequencies does not equal any of the frequencies. This makes it easier to detect exactly which tones are present in the dial signal in the presence of line distortions.

DTMF is used by most PSTN (public-switched telephone networks) systems for number dialing and is also used for voice-response systems such as telephone banking and sometimes over private radio networks to provide signaling and transferring of small amounts of data

DTMF was originally developed to allow sending control information (dialed numbers) across the telephone network. The telephone network has a bandwidth of approximately 300 to 3,400 Hz, suitable for voice communications. Any control tones would also need to be in this range and had to work regardless of whether voice was present or not. A single tone or frequency could have been used. However, if voice was present, it would interfere with the control tones, making them useless.

To overcome this, a scheme was developed whereby two tones or frequencies were combined to represent each control code or number. A total of seven tones were needed to represent the digits normally found on a telephone keypad, namely, 0–9, *, and #. These seven tones were divided into two groups of four tones each, a low-frequency group and a high-frequency group. This three-by-four array produced 12 different combinations, as shown in Figure 5.1.

You may be wondering why they chose those particular frequencies. Why not simply use multiples of say 500 Hz? The answer is in the harmonics generated due to nonlinear circuits in the phone system. If you look at the low-frequency group you will see that their second harmonic (multiple of two) falls between the high frequency tones. Third harmonics and above are outside the range of the high-frequency tones and are not a problem.

A valid tone pair has to meet the following criteria:

- Only one tone per group allowed
- Start of each tone must be less than 5 ms apart
- Both tones must last at least 40 ms
- Each tone must be within 2% of the center frequency
- The tone levels must be within 6 dB of each other

All of these features make it extremely unlikely that voice will accidentally generate valid DTMF tones.

5.3   EXPERIMENT

Now that you have an idea about the operation of a telephone, you are ready to begin the experiment.

5.3.1   DTMF Generator

In this part, you will simulate touch tone telephones, which use DTMF signals. DTMF signals are signals that consist of the sum of two pure sinusoids at valid frequencies.

To generate a DTMF signal, open LabVIEW DSP Module and create a blank VI. In the Block Diagram window, place two *Simulate Signal* Express VI located on the *Functions >> Embedded Signal Generation* palette.

Add the output of the two *Simulate Signal* VI's and connect the result to the *Analog Output*, as shown in Figure 5.2:

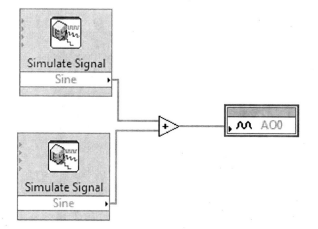

FIGURE 5.2: Adding two sinusoidal signals.

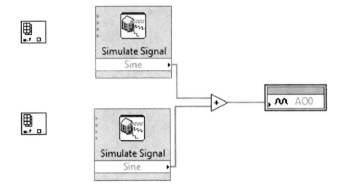

FIGURE 5.3: Adding Array Constants.

Next, the frequency of each signal needs to be specified. Add two *Index Array* VI's from the *Array* subpalette, as shown in Figure 5.3 by adding two Index Array VIs.

An array, also known as a vector or list, is one of the simplest data structures. Arrays hold equally sized data elements, generally of the same data type. Individual elements are accessed by index using a consecutive range of integers. They can be one-dimensional or *n*-dimensional. In LabVIEW, the first index in an array is 0.

Then, right-click on the array node of the two *Index Array* VI's and specify as constant.

Next, the possible frequencies should be added into the arrays. The upper array is used to generate one of the low frequency values 697, 770, 852, and 941 Hz, while the lower array will generate one of the upper frequency values 1,209, 1,336, 1,477 Hz. For example, when pressing the button "1", the dual tone generated has the two frequencies 697 and 1,209 Hz, as shown in Figure 5.5. Therefore, the element corresponding to index 1 of the upper array will hold 697, while the one in the lower array will hold 1,209. Fill in the rest of the array elements by changing the index *n*-dimensional array according to Table 5.1.

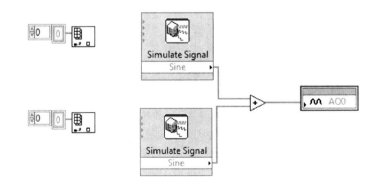

FIGURE 5.4: Adding constants to arrays.

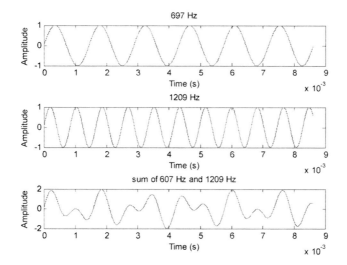

FIGURE 5.5: Adding sine waves of frequencies 607 and 1,209 Hz.

TABLE 5.1: DTMF Frequency Combinations	
DIGIT	OUTPUT FREQUENCY (Hz)
1	697+1209
2	697+1336
3	697+1477
4	770+1209
5	770+1336
6	770+1477
7	852+1209
8	852+1336
9	852+1477
0	941+1209
*	941+1336
#	941+1477

Connect the *Index Array* output to the Frequency node of the *Simulate Signal* VI.

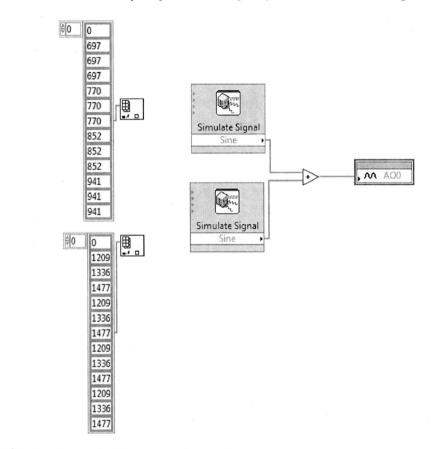

FIGURE 5.6: Connecting Frequency values to sine waves.

Next, two frequencies should be selected depending on the telephone keypad touched. To do that, in the Front Panel window go to *Boolean >> OK Button*, and place the buttons in a way to get the pattern shown in Figure 5.7.

FIGURE 5.7: Telephone keypad in Front Panel.

Go back to the Block Diagram window, and arrange the blocks for clarity.

The final step to do is to generate a dual tone by pushing one of the 12 buttons. Therefore, each button push should correspond to selecting a specific element from the two arrays to specify the two frequencies of the dual tone signal.

To do that, add a *Select* block from the *Comparison* subpalette, and connect the output of the *OK Button 6* to the *s* input of the *Select* block as shown in Figure 5.8.

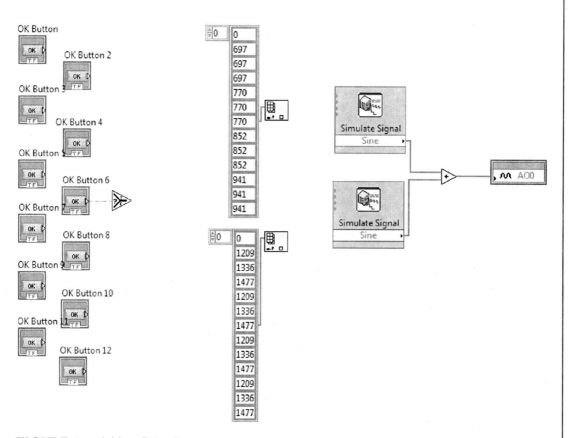

FIGURE 5.8: Adding Select Block.

If button 6 is pushed, then the elements corresponding to index 6 in the upper and lower arrays should be selected. Therefore, as shown in Figure 5.9, connect a constant of value equal to OK Button # to the *t* input of *Select* block, which corresponds to the event of pushing button 6 and a value of 0 to the *f* input.

Repeat the same for the other buttons with the corresponding values to obtain the diagram shown in Figure 5.10.

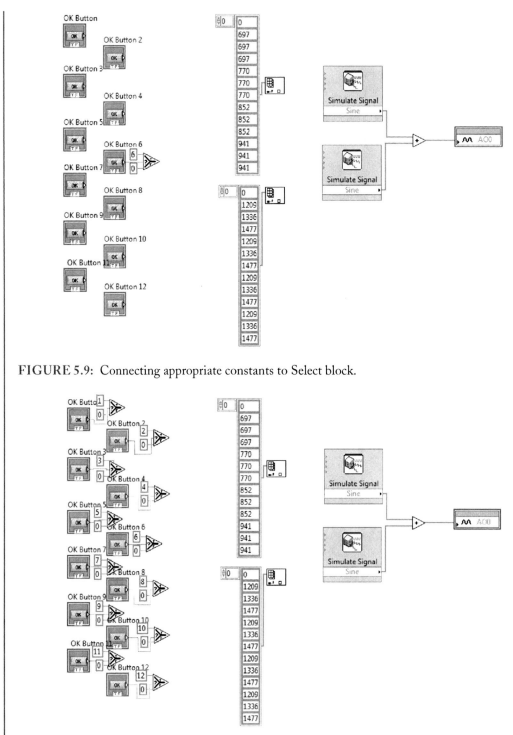

FIGURE 5.9: Connecting appropriate constants to Select block.

FIGURE 5.10: Adding constants to all the Select blocks.

Add all the outputs of the Select buttons to each other and connect the result to the index of the arrays, as follows:

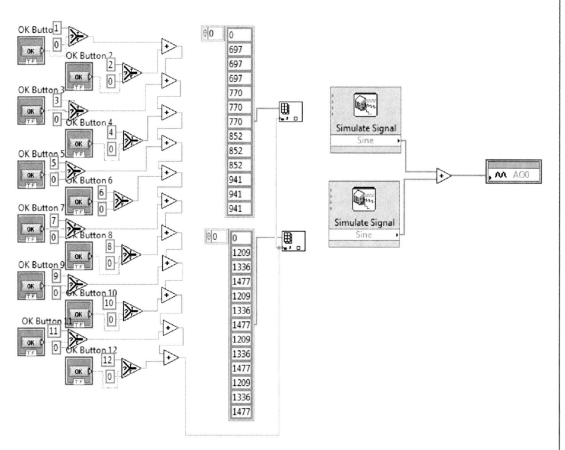

FIGURE 5.11: Adding Select blocks together.

To generate the tone for a certain period of time, one should loop the sine generation for different iterations, depending on the desired tone length. Analog Output usually determines the execution speed for the code inside the For Loop, using the following formula:

$$\frac{\text{No. samples}}{\text{Sampling rate (Hz)}} = \text{Loop speed (sec)},$$

where the number of samples and sampling rate are determined in the Simulate Signal block properties.

For example, if the number of sample is 128 and sampling rate is 8,000 Hz, then the For Loop takes 16 ms for one iteration.

Add a *For Loop* around the *Simulate Signal* blocks along with the *Analog Output* block.

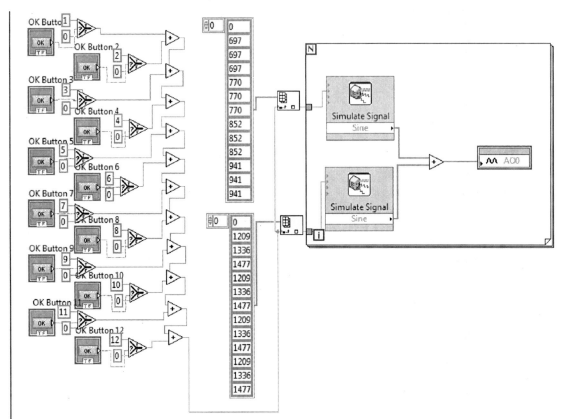

FIGURE 5.12: Adding a For Loop to generate tone for a period of time.

When dialing a telephone, the minimum tone length should be 40 ms. We will choose to generate 150 ms.

What is the number of iterations required to generate a tone of length 150 ms?

Use a *Divide* block to compute the quotient and use the result as the number of iterations as shown in Figure 5.13:

Q1

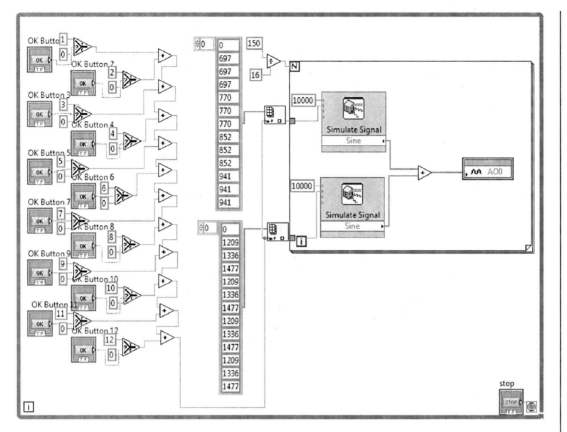

FIGURE 5.13: Final DTMF Block Diagram.

Add a *While Loop* for the VI and run it.

Try pressing different buttons and notice the tone generated with each one. Compare it with the tone generated from a touch tone phone. Comment.

Try dialing a number using the VI by pressing the corresponding buttons and holding the speaker next to the regular touch phone. Comment.

CHAPTER 6

Digital Audio Effects: Echo and Reverb

6.1 OVERVIEW

In this chapter, you will acquire a working knowledge of the basic tools used in digital audio processing. In particular, you will implement echo effects using buffers to generate delayed playback, and reverberation, which is a simulation of an echoing room.

6.2 BACKGROUND

6.2.1 Echo

Simply put, an echo takes an audio signal, and plays it back after a *delay time*. The delay time can range from several milliseconds to several seconds. Figure 6.1 presents the basic echo in a flow-graph form. This is a single echo effect, as it only produces a single copy of the input.

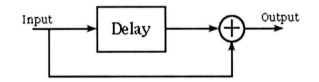

FIGURE 6.1: Diagram of the basic echo.

Just having a single echo effect is rather limiting, so most delays also have a *feedback* control (sometimes called *regeneration*) which takes the output of the delay, and sends it back to the input, as shown in Figure 6.2. Now, you have the ability to repeat the sound over and over, and it becomes quieter each time it plays back (assuming that the feedback gain is less than one. Most delay devices restrict it to be less than one for stability). With the feedback, the sound is theoretically repeated forever, but after some point, it will become so low that it will be below the ambient noise in the system and inaudible.

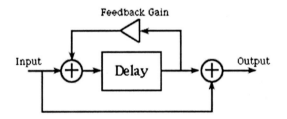

FIGURE 6.2: Diagram of echo with feedback.

6.2.2 Reverberation

Reverberation (reverb for short) is probably one of the most heavily used effects in music. When you mention reverb to a musician, many will immediately think of a stomp box, signal processor, or the reverb knob on their amplifier. But many people do not realize how important reverberation is, and that we actually hear reverb every day, without any special processors.

The effects of combining an original signal with a very short (<20 ms) time-delayed version of itself results in reverberation. In fact, when a signal is delayed for a very short time and then added to the original, the ear perceives a fused sound rather than two separate sounds. If a number of very short delays that are fed back on them are mixed together, a crude form of reverberation can be achieved.

More specifically, reverberation is the natural decay of many delayed versions of an original signal after the original has stopped sounding. It occurs in enclosed spaces as the reflections from the floors, walls, and ceiling continue to propagate after the sound source has stopped producing sound.

6.2.3 Shift Registers

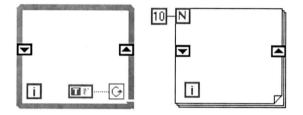

FIGURE 6.3: Shift Registers within loops in LabVIEW.

While loop and For loop structures can have terminals called shift registers that you use for passing data from the current iteration to the next iteration. Shift registers are local variables that feed forward or transfer values from the completion of one iteration to the beginning of the next.

A shift register has a pair of terminals directly opposite each other on the vertical sides of the loop border. The right terminal, the rectangle with the up arrow, stores the **data** at the completion of the iteration. LabVIEW shifts that data at the end of the iteration, and it appears in the left terminal, the rectangle with the down arrow, in time for the next iteration. You can use shift registers for any type of data, but the data you wire to each register terminals must be of the same type.

6.2.4 Echo Implementation Using Arrays

An array is one of the simplest data structures. Arrays hold a series of data elements. Individual elements are accessed by their position in the array. The position is given by an index. The index usually uses a consecutive range of integers. Some arrays are *multidimensional*, meaning they are indexed by a fixed number of integers. Generally, one- and two-dimensional arrays are the most common. The first index in LabVIEW is 0.

Although a buffer for delay would consist of several thousand elements, let us consider a simple buffer containing just eight elements.

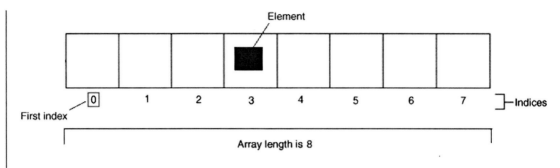

Q1   FIGURE 6.4: Array structure.

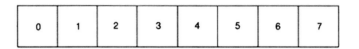

FIGURE 6.5: Array with eight elements.

The elements within the buffer are numbered from 0 to 7. The first element is buffer[0], the second element buffer[1], and the final element is buffer[7]. Zeroes are assigned as initial values to each element. One way to use the buffer as a delay is to save the newest measurement in one end of the buffer and the oldest at the other. For example, we may wish to use buffer[0] to hold the newest input and buffer[7] to hold the oldest. This arrangement is known as a *straight buffer*. Every time a measurement is made by the input, a new value is put into buffer[0]. We discard the oldest reading at buffer[7] then shuffle all the values along one place. For example, buffer[6] would be moved into buffer[7]. Diagrammatically, the process is shown in Figure 6.6, assuming that the buffer already contains the values 1, 2, 3, 4, 5, 6, 7, and 8.

FIGURE 6.6: Straight buffer arrangement.

Unfortunately, this arrangement is not suitable for implementing echo and reverberation. If the buffer is large, say several thousand values, most of the time is spent by the program moving data from one location to another. This does not leave enough time to do the audio processing! A more suitable arrangement is to use what is known as a *circular buffer*, which will be covered in this chapter.

A circular buffer works in a different way. Values within the buffer are not shuffled along. Instead, the oldest element is overwritten by the newest. This means that the location of the oldest element in the buffer moves every time a new sample is saved. We keep track of the position of the oldest element with a pointer as shown in Figure 6.7, which assumes that the buffer already contains the values 1, 2, 3, 4, 5, 6, 7, and 8.

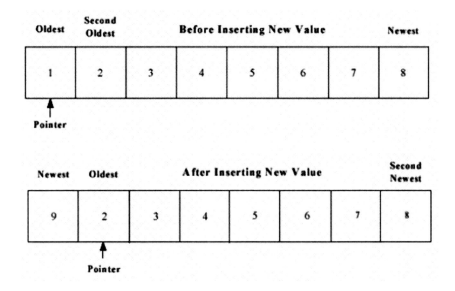

FIGURE 6.7: Circular buffer arrangement.

To insert a new value into the buffer, the oldest value is simply overwritten and the pointer moved along one place. Unlike a straight buffer, all the other values still remain in exactly the same place—they are not shuffled along.

When the end of the buffer is reached, the oldest value is overwritten with the newest. In this case, the pointer is set back to the beginning of the buffer again, as if the buffer were circular.

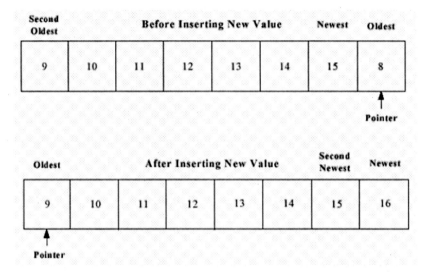

FIGURE 6.8: Setting back pointer to beginning of buffer.

6.3 EXPERIMENT

Now that you have an idea about the echo and reverb effects, you are ready to begin the experiment and implement them.

6.3.1 Application 1: Echo

As a starting application, you will implement an echo. The echo is one of the simplest effects out there, but it is very valuable when used properly. A little echo can bring life to dull mixes, widen your instrument's sound, and even allow you to solo over yourself.

The first thing to do is to replicate in LabVIEW the diagram shown in Figure 6.1.

The Input and Output blocks are selected from the *Functions* palette by choosing the *Analog Input* and *Analog Output* nodes, respectively, found in *Elemental I/O*.

Configure these blocks to have a framesize of 1, 8,000 Hz sampling rate, and a gain of 1.

Next, the Sum block should be added by selecting, from the *Functions* palette, the *Sum* node found in the *Numeric* subpalette.

Connect the output of the Analog Input to the input lines of the Sum block, and the output line of the Sum block to the input of the Analog Output.

The resulting diagram should look as shown in Figure 6.9:

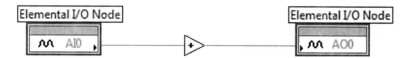

FIGURE 6.9: Echo implementation step 1.

Q2

Next, the Delay block should be implemented which delays the input by a specific amount of time and adds it to the second input of the Sum block.

To do so, insert an array which will be used to store past inputs. This is done by choosing from the *Array* subpalette ⊞ the *Index Array* ⊞.

⊞

FIGURE 6.10: Inserting array to store past inputs.

Q3

The next step is to replace the oldest value stored in the array by the newest data acquired from the *Analog Input* and shifting the array data by one place.

Therefore, add *Replace Array Subset* ⊞ from the *Array* subpalette ⊞. This block will be used to replace the oldest sample stored in the array by the newest sample acquired from the *Analog Input*. Therefore, connect the output of the *Analog Input* to the new element node of *Replace Array Subset* as shown in Figure 6.11.

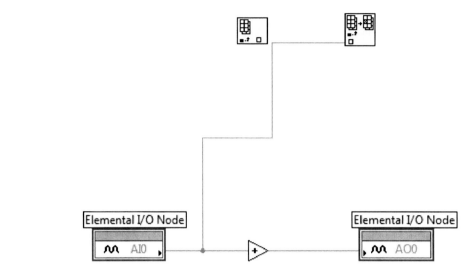

FIGURE 6.11: Analog Input connected to Replace Array Subset.

However, the user should be able to control past inputs and attenuate them. Therefore, add a Multiply block and connect a control slider and the output of the Analog Input to the two inputs of the Multiply block.

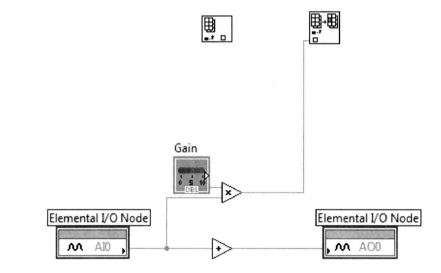

FIGURE 6.12: Controlling past inputs' attenuation using a slider.

The next step is to choose the oldest value from the array and add it to the newest value acquired from the Analog Input so that an echo is generated.

Therefore, connect the output of the *Index Array* to the second input of the *Add* block as shown in Figure 6.13.

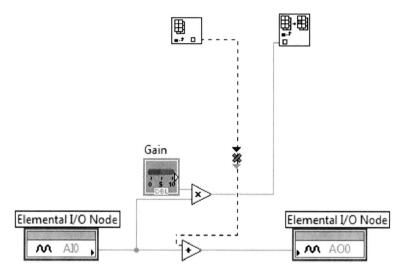

FIGURE 6.13: Adding oldest data to newest data to generate echo.

The next step is to increment the pointer to the oldest value as in the circular buffer method explained in the Background section.

To do so, add the *Quotient and Remainder* block from the *Numeric* subpalette. Connect its $x-y*$floor(x/y) node to the index of the *Index Array* block and *Replace Array Subset* as shown in Figure 6.14.

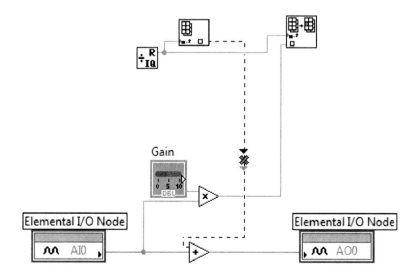

FIGURE 6.14: Connecting Quotient and Remainder block to Index Array.

Add a slider to the *y* input of the *Quotient and Remainder* block, which allows you to control the delay, as shown in Figure 6.15. Make sure that the delay range is 0–1,000.

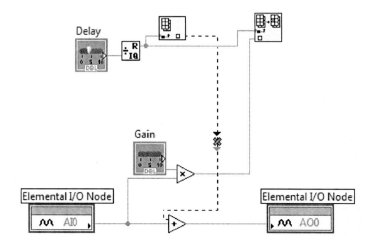

FIGURE 6.15: Adding Delay slider.

Knowing that the delay in seconds is calculated using the sampling rate and the Delay slide as Delay (seconds) = Delay (samples)/Sampling Rate (samples per second), what is the maximum delay in seconds that can be achieved?

To increase the pointer to the oldest data, add an *Increment* block ▷ from the *Numeric* subpalette ▷.

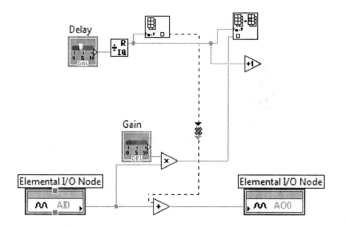

FIGURE 6.16: Incrementing the pointer.

Because the pointer is incremented in each loop, add a *While loop* to all the blocks.

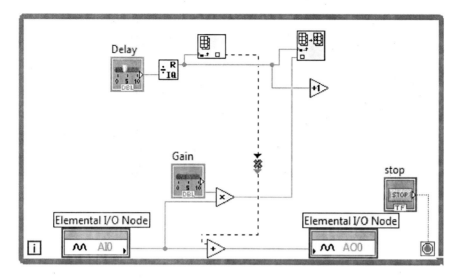

FIGURE 6.17: Adding a While loop.

The last step is to shift the values in the buffer circularly. This is done by simply right-clicking on the border of the While Loop and choosing *Add Shift Register*, as shown in Figure 6.18.

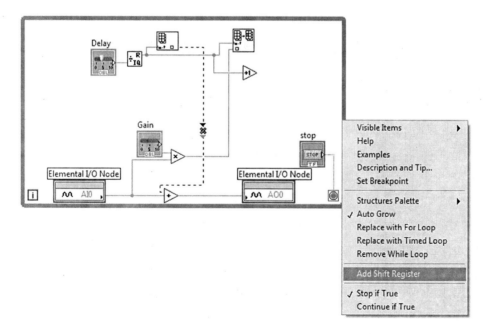

FIGURE 6.18: Adding shift Register.

Connect the output of the Replace Array Subset to the right terminal, the rectangle with the up arrow. It stores the data at the completion of the iteration. LabVIEW shifts that data at the end of the iteration, and it appears in the left terminal, the rectangle with the down arrow, in time for the next iteration. Connect the latter to the array node of the Index Array, as shown in Figure 6.19.

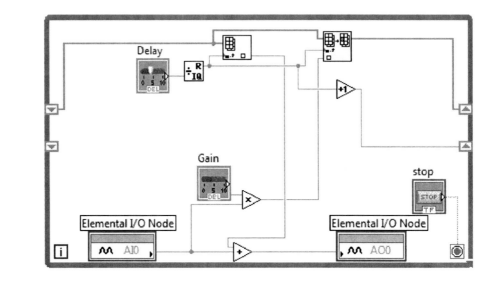

FIGURE 6.19: Shifting values of arrays.

Repeat the same with output of the Increment block as shown in Figure 6.20.

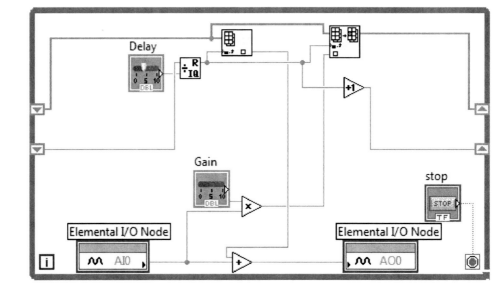

FIGURE 6.20: Shifting values of pointer.

Initialize the upper shift register to an array of zeroes and the lower one to zero.

To initialize a shift register, wire a value from outside the loop to the left terminal. If you do not initialize the register, the loop uses, as the initial value, the last value inserted in the register when the loop last executed, or the default value for its data type if the loop has never before executed. You should normally use initialized shift registers to ensure consistent behavior.

The array should initially contain 1,000 elements of zeroes. Therefore, add a *Simulate Signal* VI and double-click on it. Configure the VI to have *DC* signal type, Offset equal to 0, 8,000 samples per second, and 1,000 samples, as shown in Figure 6.21.

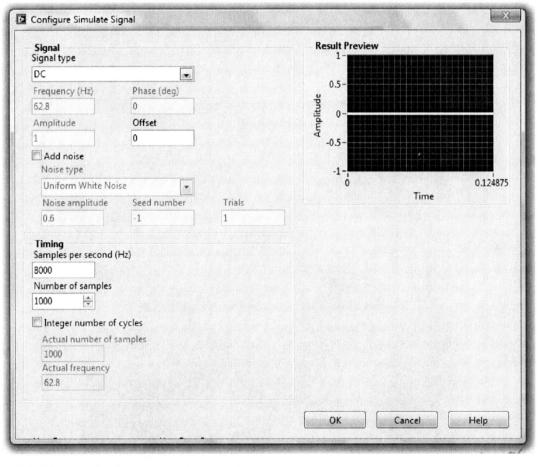

FIGURE 6.21: Configuring Simulate Signal.

Connect the output of the *Simulate Signal* VI to the upper left shift register and a constant 0 to the lower shift register as shown in Figure 16.

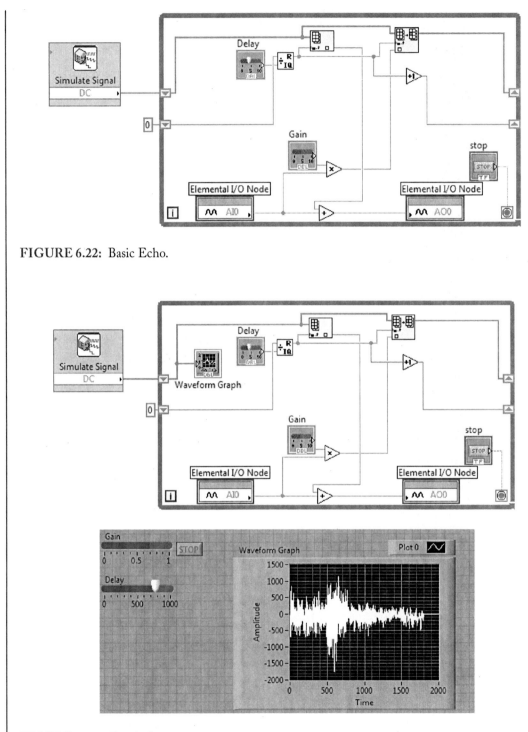

FIGURE 6.22: Basic Echo.

FIGURE 6.23: Graph for past inputs.

Add a waveform graph at the input of the array to see the past inputs as shown in Figure 6.23.

Run the VI and try playing with the delay. Notice the effect on the Echo. Comment.

You can make the echo more advanced by adding the oldest input and the newest input from the Analog Input and storing them as the newest input, as explained in the Background section. The result is shown in Figure 6.24. Make sure the Gain range is between 0 and 1.

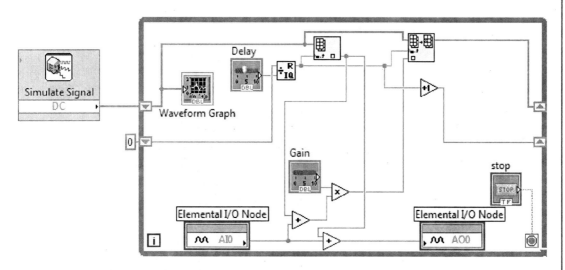

FIGURE 6.24: Echo with feedback.

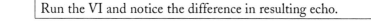

Run the VI and notice the difference in resulting echo.

6.3.2 Application 2: Reverb

Modify the diagram as shown in Figure 6.25 and change the framesize of the Analog Input to 128 and the delay range to 0–127.

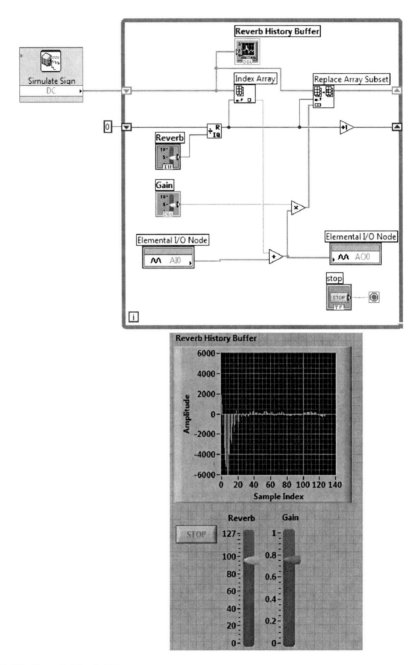

FIGURE 6.25: Reverb Block Diagram.

Run the VI and compare with Echo effect.

CHAPTER 7

Music Composer

7.1 OVERVIEW

This chapter demonstrates the capability of LabVIEW to generate musical tones using a DSP board. You will simulate a piano by pressing the keys on the front panel to generate any combination of musical notes on your speaker.

7.2 BACKGROUND
7.2.1 Creating Digital Music

When a sine wave is output to a speaker. one will hear a continuous tone on the speaker. The *loudness* of the tone is determined by the amplitude of the wave. The *pitch* is defined as the frequency of the wave.

The table below contains frequency values for the notes in one octave.

TABLE 7.1: Frequencies corresponding to musical notes	
NOTE	**FREQUENCY (HZ)**
A	$220 = 220 \times 2^{0/12}$
A#	$233 = 220 \times 2^{1/12}$
B	$247 = 220 \times 2^{2/12}$
C	$262 = 220 \times 2^{3/12}$
C#	$277 = 220 \times 2^{4/12}$
D	$294 = 220 \times 2^{5/12}$
D#	$311 = 220 \times 2^{6/12}$
E	$330 = 220 \times 2^{7/12}$
F	$349 = 220 \times 2^{8/12}$
F#	$370 = 220 \times 2^{9/12}$
G	$392 = 220 \times 2^{10/12}$
G#	$415 = 220 \times 2^{11/12}$
A	$440 = 220 \times 2^{12/12}$

The frequency of each note can be calculated by multiplying the adjacent previous note frequency by $\sqrt[12]{2}$. You can use this method to determine the frequencies of additional notes above and below the ones in Table 7.1. There are 12 notes in an octave. therefore moving up one octave doubles the frequency. A sequence of notes can be separated by pauses (silences) so that each note is heard separately. The *envelope* of the note defines the amplitude versus time.

A chord is created by playing multiple notes simultaneously. When two piano keys are struck simultaneously both notes are created. and the sounds are mixed arithmetically. You can create the same effect by adding two sinusoidal waves together in software. before sending the wave to the DAC. Figure 7.1 plots the mathematical addition of a 262-Hz (low C) and a 392-Hz sine wave (G). creating a simple chord.

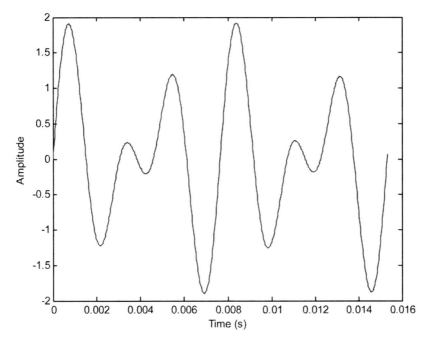

FIGURE 7.1: Chord tone of low C and G.

7.3 EXPERIMENT

In this experiment. you will use the Sine Waveform generation in LabVIEW to generate musical tones using a DSP board.

7.3.1 Music Composer

In the Front Panel, insert *OK button* from the *Boolean* subpalette.

FIGURE 7.2: OK button.

Single-click on the button to highlight it.

FIGURE 7.3: Highlighting OK button.

Drag the borders to shape the button as a piano key:

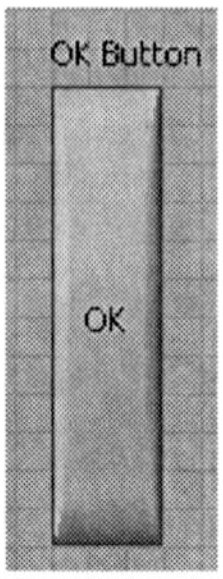

FIGURE 7.4: Shaping OK button as piano key.

Right-click on the button and go to *Properties*.
Modify the *Properties* window as shown in Figure 7.5:

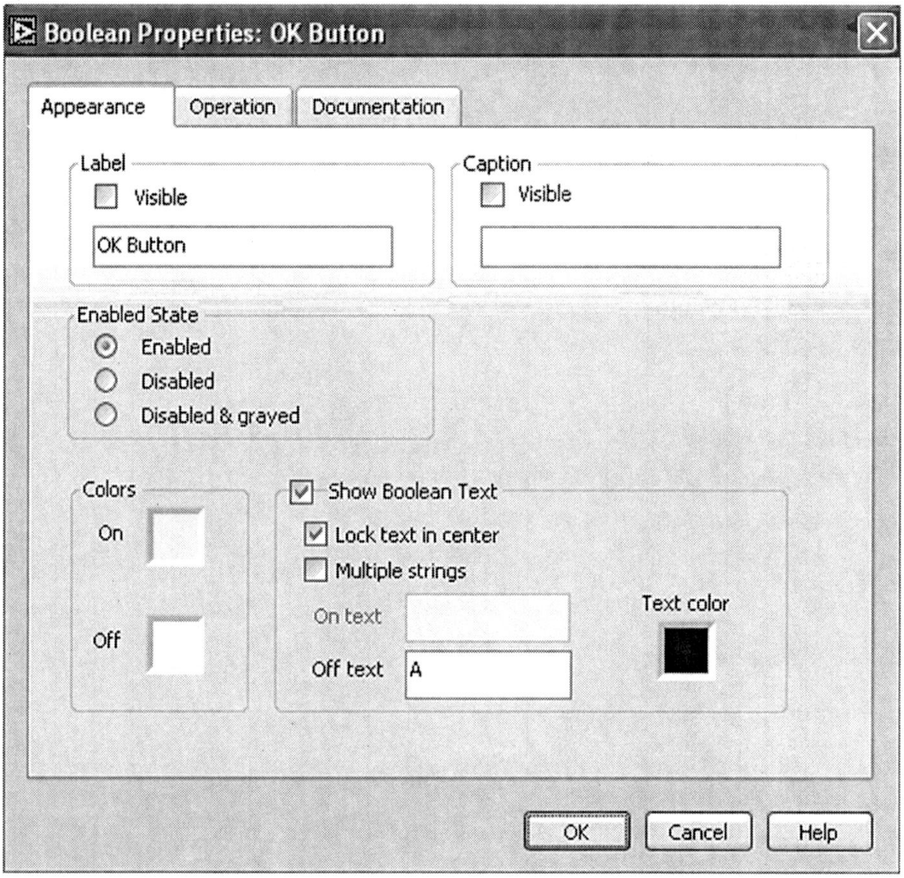

FIGURE 7.5: OK button properties.

The button should look as follows after modifying the properties window:

FIGURE 7.6: OK button after modifying properties.

Insert six other Boolean buttons and repeat the process. to obtain the following:

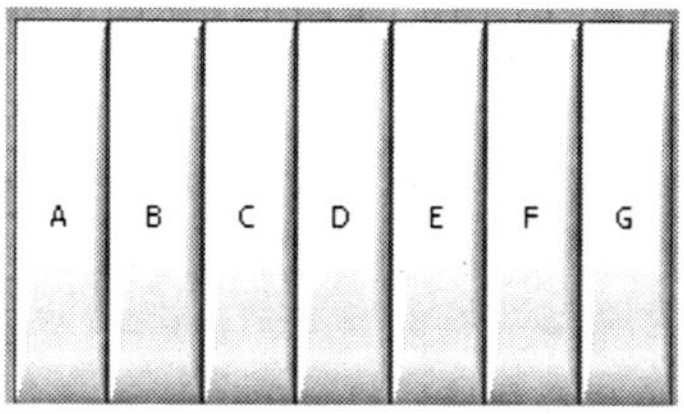

FIGURE 7.7: Piano keyboard.

These buttons represent the piano keys. Pressing a button should generate a corresponding frequency representing the musical note.

In the Block Diagram. you should have the following:

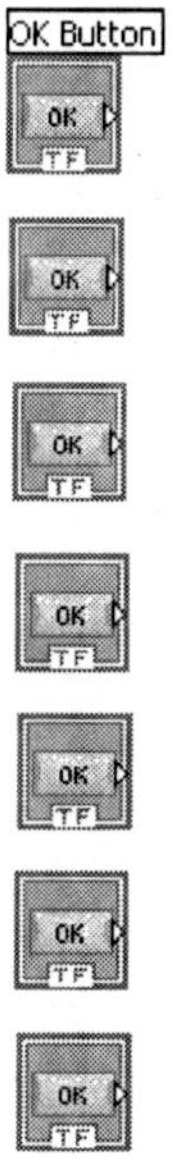

FIGURE 7.8: OK buttons in block diagram.

Add seven *EMB Sine Waveform* blocks (EMB stands for Extended Memory Block) corresponding to each button as shown in Figure 7.9:

FIGURE 7.9: Adding EMB Sine Waveforms.

The frequency of each sine wave should be modified to correspond with each note. as specified by the table in the *Background* section. Modify the Block Diagram to obtain the diagram shown in Figure 7.10:

One should also be able to control the loudness of the tone. This can be done by adding a control knob to control the amplitude node of the EMB sine wave block as shown in Figure 7.11:

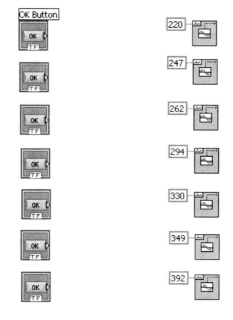

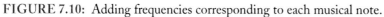

FIGURE 7.10: Adding frequencies corresponding to each musical note.

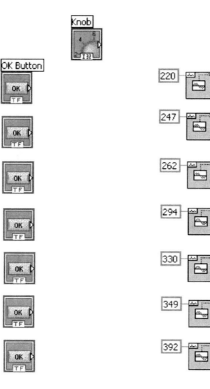

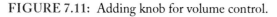

FIGURE 7.11: Adding knob for volume control.

Make sure the amplitude range of the knob is between 0 and 2000.

When a piano key is not pushed. the amplitude of the corresponding note should be 0. Therefore. modify the block diagram by inserting constants a 0 constant and 7 selection functions from Comparison subpalette as shown in Figure 7.12:

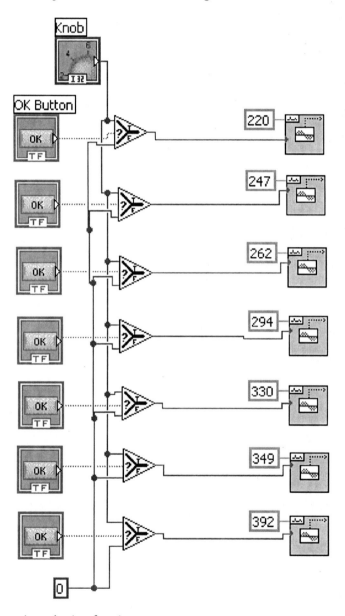

FIGURE 7.12: Inserting selection functions.

These seven selector functions select either 0 (off) or Amplitude (on). which will be used to determine the amplitude of the sine wave produced.

Now. all of the sine waves should be added together to create a waveform which will contain all of the selected notes. as shown in Figure 7.13.

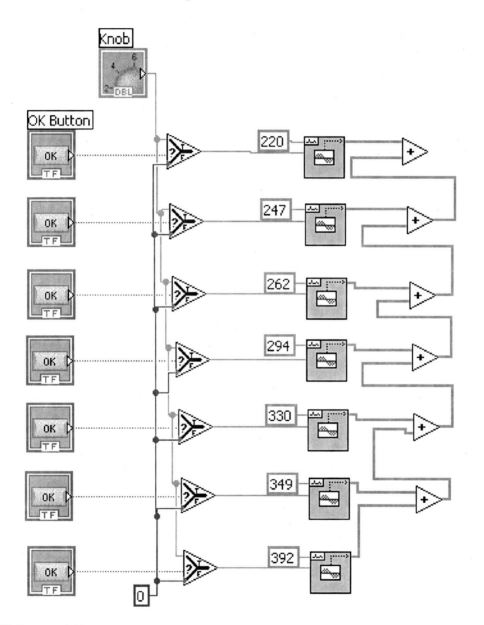

FIGURE 7.13: Adding sine waves together.

The last step is to output the tones to the audio jacks on the DSP board. Therefore. add an Analog Output and configure it to have two channel samples and an 8.000-Hz sampling frequency. Connect the output of the Add block to the Analog Output.

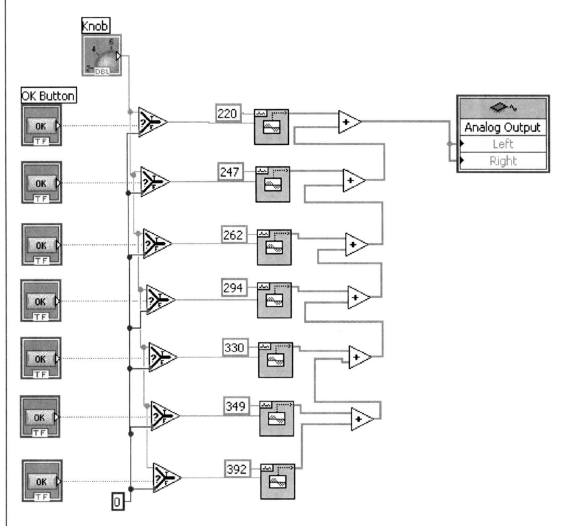

FIGURE 7.14: Adding analog output.

Add a While loop to the whole Block Diagram and run the VI.

Try to press the piano keys. Can you hear any sound?
Knowing that by default the EMB sinewave has 128 samples and 8.000 samples/s. How long will each tone be generated for?

To make the tone last longer. connect the number of samples (#s) node to a constant of 1,500. by right-clicking on the #s node and creating a constant.

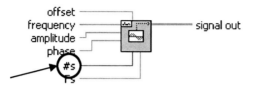

FIGURE 7.15: #s node in EMB sine waveform.

Run the VI again and notice the tone sound when pressing different keys. How long does the tone last for?

The piano has actually 88 keys which span the frequency range 27.5 Hz (A0) to 4.186 Hz (C8). Table 7.2 below shows the 88 keys and their frequencies.

TABLE 7.2: Piano keys and their frequencies		
KEY NUMBER	NOTE NAMES	FREQUENCY (HZ)
88	C8	4,186.01
87	B7	3,951.07
86	A#7/Bb7	3,729.31
85	A7	3,520.00
84	G#7/Ab7	3,322.44
83	G7	3,135.96
82	F#7/Gb7	2,959.96
81	F7	2,793.83
80	E7	2,637.02

TABLE 7.2: (Continued)		
KEY NUMBER	NOTE NAMES	FREQUENCY (HZ)
79	D#7/Eb7	2,489.02
78	D7	2,349.32
77	C#7/Db7	2,217.46
76	C7	2,093.00
75	B6	1,975.53
74	A#6/Bb6	1,864.66
73	A6	1,760.00
72	G#6/Ab6	1,661.22
71	G6	1,567.98
70	F#6/Gb6	1,479.98
69	F6	1,396.91
68	E6	1,318.51
67	D#6/Eb6	1,244.51
66	D6	1,174.66
65	C#6/Db6	1,108.73
64	C6	1,046.50
63	B5	987.767
62	A#5/Bb5	932.328
61	A5	880.000
60	G#5/Ab5	830.609
59	G5	783.991

TABLE 7.2: (*Continued*)		
KEY NUMBER	NOTE NAMES	FREQUENCY (HZ)
58	F#5/Gb5	739.989
57	F5	698.456
56	E5	659.255
54	D#5/Eb5	622.254
54	D5	587.330
53	C#5/Db5	554.365
52	C5	523.251
51	B4	493.883
50	A#4/Bb4	466.164
49	A4 concert tone	440.000
48	G#4/Ab4	415.305
47	G4	391.995
46	F#4/Gb4	369.994
45	F4	349.228
44	E4	329.628
43	D#4/Eb4	311.127
42	D4	293.665
41	C#4/Db4	277.183
40	C4 (middle C)	261.626
39	B3	246.942
38	A#3/Bb3	233.082

TABLE 7.2: (*Continued*)		
KEY NUMBER	NOTE NAMES	FREQUENCY (HZ)
37	A3	220.000
36	G#3/Ab3	207.652
35	G3	195.998
34	F#3/Gb3	184.997
33	F3	174.614
32	E3	164.814
31	D#3/Eb3	155.563
30	D3	146.832
29	C#3/Db3	138.591
28	C3	130.813
27	B2	123.471
26	A#2/Bb2	116.541
25	A2	110.000
24	G#2/Ab2	103.826
23	G2	97.9989
22	F#2/Gb2	92.4986
21	F2	87.3071
20	E2	82.4069
19	D#2/Eb2	77.7817
18	D2	73.4162
17	C#2/Db2	69.2957

| \multicolumn{3}{c}{TABLE 7.2: (*Continued*)} |
KEY NUMBER	NOTE NAMES	FREQUENCY (HZ)
16	C2	65.4064
15	B1	61.7354
14	A#1/Bb1	58.2705
13	A1	55.0000
12	G#1/Ab1	51.9130
11	G1	48.9995
10	F#1/Gb1	46.2493
9	F1	43.6536
8	E1	41.2035
7	D#1/Eb1	38.8909
6	D1	36.7081
5	C#1/Db1	34.6479
4	C1	32.7032
3	B0	30.8677
2	A#0/Bb0	29.1353
1	A0	27.5000

Given the above frequencies, simulate a complete piano using the SPEEDY-33 board.

CHAPTER 8

Introduction to Robotics

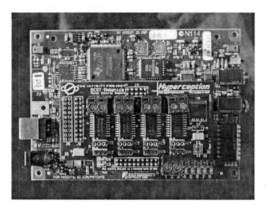

8.1   OVERVIEW

This chapter provides an introduction on how to construct a robotic device using the SPEEDY-33 board and the NI LabVIEW Embedded DSP module software.

8.2   BACKGROUND

8.2.1   SPEEDY-33 With Robotic Applications

One of the reasons the NI SPEEDY-33 is such an excellent educational tool is because of its wide range of functionality. As we will explore in this chapter, one such capability of the board is as a robotics control unit [1]. Using an auxiliary robotics daughter module, which attaches directly to the SPEEDY-33, as well as the existing board capabilities and LabVIEW DSP software, we will create a stand-alone robotics application that is able to move and navigate obstacles autonomously.

The robotics daughter module attachment for the SPEEDY-33 provides connectivity for several components. There are four outputs for DC drive motors, which are used to provide the primary motion for our robotics application. There are also four outputs for servo motors, which are small motors with built-in control circuitry that allows them to be programmed to operate at specific angular positions and can be used for additional motion (i.e., adding auxiliary components such as a lift, drill, rotating disco ball, etc.). In addition, there are four RC connections, which can be connected to various control devices, such as microcontrollers or bump sensors, which we will use in this particular application. When paired with the DSP chip and other components of the "motherboard" SPEEDY-33, as well as our existing knowledge of programming using LabVIEW, the robot daughter module provides us with all the tools necessary to create our machine.

However, there are some constraints we must work within in the development of our robot. As previously discovered, the memory capability of the SPEEDY-33 is somewhat limited. We have only 2,048 kb of flash memory at our disposal, which is responsible for holding every piece of information required to control our stand-alone creation. While sufficient for a simple driving program, this creates a challenge as we begin to add more and more functionality to the robot. As we increase the size of our control program, as well as explore additional applications that typically require large amounts of data storage (think Mars Rover, photography, etc.), we meet severe memory limitations which must be accounted for during project design. In addition, drive motor, servo, and RC inputs/outputs are limited, which places a ceiling on the mechanical capability of our robot. It is possible to circumvent mechanical limitations with additional programming, but that, in turn, places a strain on our already limited memory supply. These are all factors to be considered to ensure a highest-quality result.

Because of the wide range of functionality provided by LabVIEW and the SPEEDY-33, there are several prior projects that have influenced and contributed to our particular design. Most notably, students at Texas A&M University developed a pickup truck-like robot using the SPEEDY-33 and used it for various applications, similar to the project we are undertaking [2]. They allowed their device to run either online with the computer or autonomously, and used the SPEEDY-33 and a buffer board to act as the "brains" of the vehicle. They used a rechargeable battery system with a regulating board under the "hood" of the vehicle to simulate the power plant and alternator system common in automobiles today. Finally, they added a connected breadboard in the "bed" of their truck to allow the user to add "loads" to the vehicle and evaluate their affect on performance. The Infinity Project, created by Southern Methodist University, Santa Clara University, and Texas Instruments, uses LabVIEW and the SPEEDY-33 to immerse robotics and programming into high school curriculum, and allows students to complete a variety of different projects with different applications [3]. Finally, the SPEEDY-33 was used by a group of students at Texas Tech University to create a machine that competed in the annual West Texas BEST Robotics Competition [4]. The students created a manually controlled device that acted as a ball retrieval system in an arena-based automated ball game. These are all excellent examples that can be used for inspiration or guidance in the completion of your robotics project.

8.2.2   SPEEDY-33/ Robot Daughter Module Overview

By now, we should all be familiar with the SPEEDY-33 board. However, in this chapter, we will expand upon the existing capabilities of the SPEEDY-33 with the auxiliary robotics daughter module. Figure 8.1 below shows a physical layout of the SPEEDY-33.

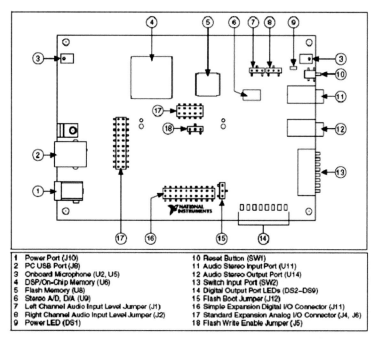

1 Power Port (J10)	10 Reset Button (SW1)
2 PC USB Port (J8)	11 Audio Stereo Input Port (U11)
3 Onboard Microphone (U2, U5)	12 Audio Stereo Output Port (U14)
4 DSP/On-Chip Memory (U6)	13 Switch Input Port (SW2)
5 Flash Memory (U8)	14 Digital Output Port LEDs (DS2–DS9)
6 Stereo A/D, D/A (U9)	15 Flash Boot Jumper (J12)
7 Left Channel Audio Input Level Jumper (J1)	16 Simple Expansion Digital I/O Connector (J11)
8 Right Channel Audio Input Level Jumper (J2)	17 Standard Expansion Analog I/O Connector (J4, J6)
9 Power LED (DS1)	18 Flash Write Enable Jumper (J5)

FIGURE 8.1: SPEEDY-33 layout.

Take note of items 16 and 17 on the board layout. They are the Simple Expansion Digital I/O Connector and Standard Expansion Analog I/O Connector, respectively. These are the components that the robotics daughter module will attach to.

Now observe a layout diagram for the Daughter Module, shown in Figure 8.2 below.

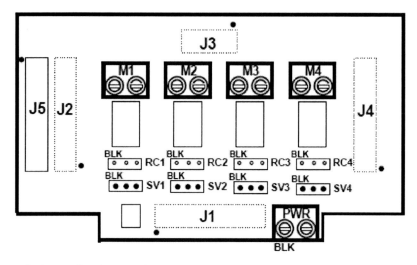

FIGURE 8.2: Robotics daughter module layout.

The major relevant components are as follows:

- M1-M4: Main/Drive Motor Outputs
- SV1-SV4: Auxiliary Servo Motor Outputs
- RC1-RC4: R/C or other Controller Inputs
- PWR: Power Connection

Turn over your robotics daughter module and observe the underside of the board. As you can see, there are several pins that align directly with the Digital and Analog Expansion Connectors on the SPEEDY-33. If positioned correctly, the Daughter Module will attach directly to the top of the SPEEDY-33. Figure 8.3 below shows a picture of the SPEEDY-33 with the robotics daughter module attached.

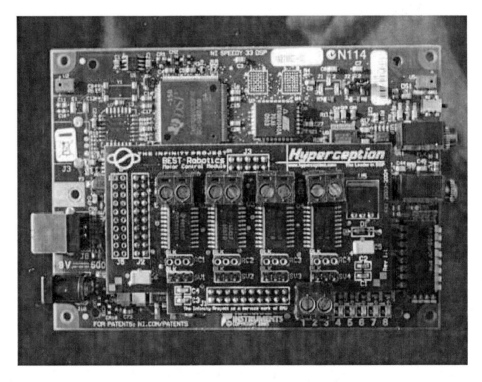

FIGURE 8.3: SPEEDY-33 with robotics daughter module attached.

8.3   EQUIPMENT/TOOLS

The following is an overview of most of the equipment and tools needed to complete the SPEEDY-33 Robotics Project [5]:

PHYSICAL STRUCTURE:

Body Material. *Note: Clear Plexiglas can be used to allow students to easily view the internal components of the robot. However, any rigid material will do. Important things to keep in mind are strength, flexibility, weight, cost, ease of shaping/ cutting, etc.*

Adhesives, brackets, etc for proper construction of body

Necessary tools for shaping/construction of body (i.e., epoxy, drill, saw, Plexiglas cutting tool, screws, nuts, bolts, etc.) *Note: Please DO NOT purchase a brand new tool set solely for the purpose of this project! Find tools to borrow, or inquire as to the use of ASU's machine shop for larger tools such as drills, saws, etc.*

Wheels/Axles. *Note: keep in mind size of wheels in relation to the rest of your robot and how that will affect functionality, as well as the connection between the motor shafts and your wheels.*

SOFTWARE/PROCESSING EQUIPMENT:

NI SPEEDY-33 Board with attached robotics daughter module

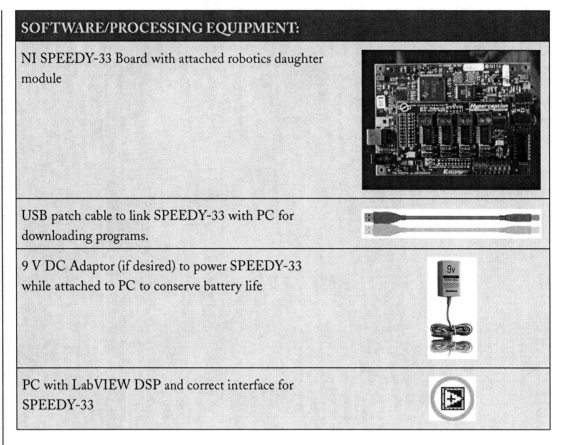

USB patch cable to link SPEEDY-33 with PC for downloading programs.

9 V DC Adaptor (if desired) to power SPEEDY-33 while attached to PC to conserve battery life

PC with LabVIEW DSP and correct interface for SPEEDY-33

ELECTRICAL HARDWARE:

Two 9 V Drive Motors. *Note: Be aware of how motors will attach to the body of your robot, as well as how the drive shafts will connect to your drive wheels.*

Two Bump Sensors	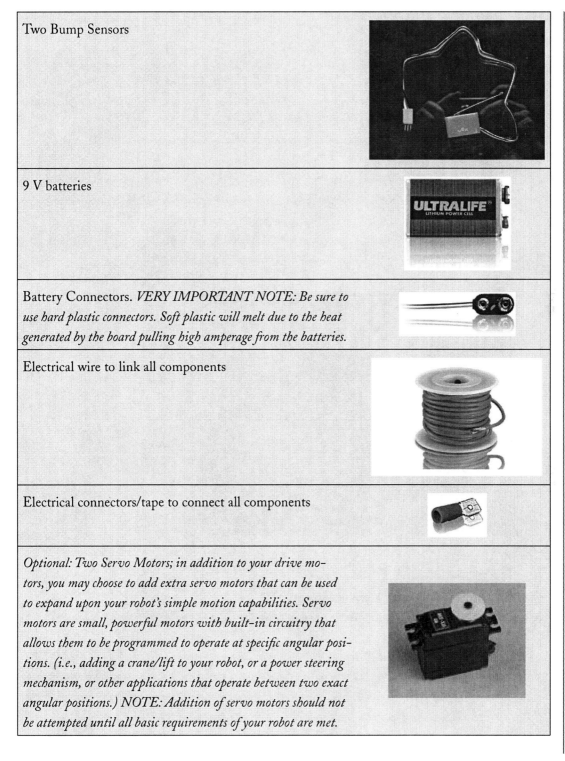
9 V batteries	
Battery Connectors. *VERY IMPORTANT NOTE: Be sure to use hard plastic connectors. Soft plastic will melt due to the heat generated by the board pulling high amperage from the batteries.*	
Electrical wire to link all components	
Electrical connectors/tape to connect all components	
Optional: Two Servo Motors; in addition to your drive motors, you may choose to add extra servo motors that can be used to expand upon your robot's simple motion capabilities. Servo motors are small, powerful motors with built-in circuitry that allows them to be programmed to operate at specific angular positions. (i.e., adding a crane/lift to your robot, or a power steering mechanism, or other applications that operate between two exact angular positions.) NOTE: Addition of servo motors should not be attempted until all basic requirements of your robot are met.	

8.4   EXPERIMENT

8.4.1   Bump Sensor Activation

I.   Connecting Bump Sensors

    a.   The robot daughter module contains four RC Inputs. Locate the RC inputs on the robot daughter module. The inputs are normally connected to a microcontroller, but in here, bump sensors will be attached in their place for autonomous capability.

    b.   The channel number is next to each input. In the test program, only RC Inputs 0 and 1 are used, so connect your bump sensors to those two inputs.

    c.   The bump sensors are composed of two wires. Make sure the white wire is connected to Pin1 of the RC input connector. This pin is recognized by the "BLK" label written above Pin1 of each RC input. The black wire should be connected to Pin3.

    d.   For RC inputs and servo motors, an external battery source is not required. You can test the bump sensors with the standard USB power connection.

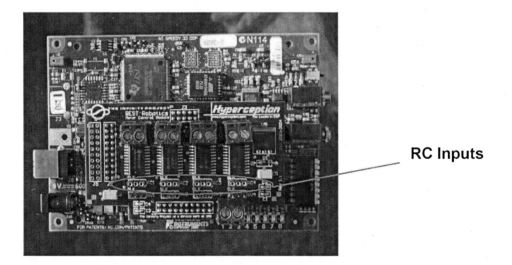

RC Inputs

FIGURE 8.4: RC inputs.

II.   Building Test Program for Bump Sensors

    a.   Open a blank VI and insert a while loop from the *Functions* palette to enclose the program.

    b.   Select two *RC Speed Control VIs*, as shown below, from the *Functions* → *SPEEDY-33* drop down menu. The RC Speed Control is normally used to gain speed and direction data from a R/C microcontroller. For this project, you will be using these devices to detect voltage changes brought about by your bump sensors.

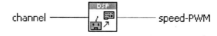

channel ———— speed-PWM

RC Speed Control.vi

FIGURE 8.5: LabVIEW bump sensor input.

   c.   Specify the receiver input channel for the RC SPEED Control by right-clicking on the VI and selecting *Create → Constant → RC1*. Another option is to double-click on the RC Speed Control VI and select the RC input from the Channel control. Designate your second bump sensor as input *RC2*.

   d.   Create two waveform charts and connect them to the outputs of the RC Speed Control VIs as displayed below.

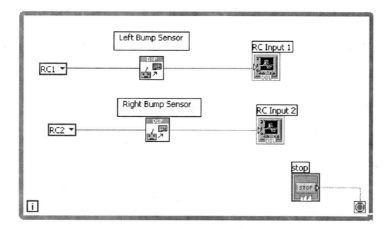

FIGURE 8.6: LabVIEW complete bump sensor program.

   e.   The test program is complete. Save this program, as it will be used later to create the Bump Sensor Detection VI.

III. Running the Bump Sensor Test Program

   a.   Make sure that the digital input (DIP) switches 1 and 2 are not pressed. The activation of DIP switch 1 turns off RC1, while DIP switch 2 deactivates RC2.

   b.   Load the Bump Sensor Test Program on the SPEEDY-33 Board.

   c.   When you press a bump sensor, the corresponding waveform chart should read an output value of -1 V. The screenshot of the waveform chart below demonstrate the expected values.

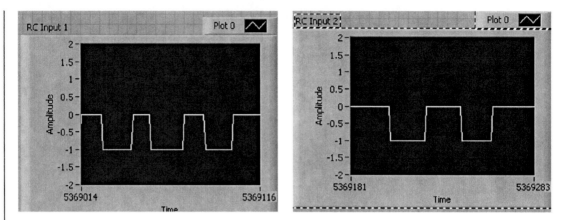

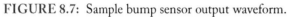

FIGURE 8.7: Sample bump sensor output waveform.

8.4.2   Motor Activation

I.   Connecting DC Motors

    a.   The robot daughter module includes four motor drive outputs (M1, M2, M3, and M4). In this experiment, only two motor drive outputs M1 and M2 are used.

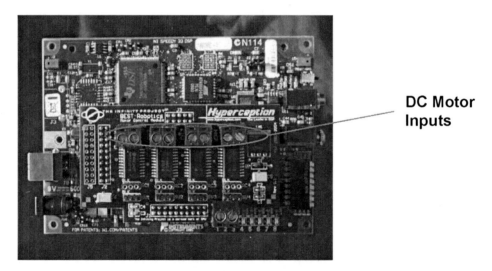

DC Motor Inputs

FIGURE 8.8: DC motor outputs.

    b.   Unscrew the terminal blocks and insert the wires connected to the DC motors. Connect the DC motors to *CH1* and *CH2*.

    c.   To run the DC motors, an external 9 V battery must be connected to the robot daughter modules external battery source:

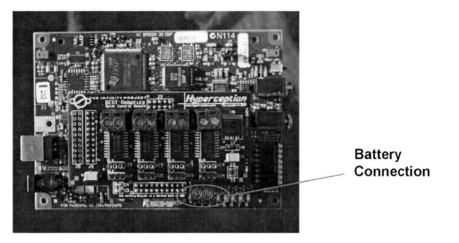

FIGURE 8.9: External battery source connection.

    d.   *Warning*: When connecting the 9-V battery, do not use a thin plastic battery cover; the battery will not run the motors and will overheat very quickly. Instead use a thick plastic battery cover.

    e.   *Do not connect the battery, until the program has been downloaded to the device.* Unless a program has sent a specified speed to the DC motors, they will run at full speed. Therefore, wait until the program has been loaded to connect the battery.

II. Building Test Program for DC Motors

    a.   Open a blank VI and insert a while loop to enclose the program.

    b.   Select two *Motor Drive VIs*, as shown below, from the *Functions → SPEEDY-33* drop down menu. The Motor drive VI receives a speed value from an R/C microcontroller or $_{user}$ input to control a motor at a designated output.

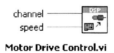

Motor Drive Control.vi

FIGURE 8.10: LabVIEW drive motor control.

    c.   Specify the receiver input channel for the Motor Drive output by right-clicking on the VI and selecting *Create → Constant → Control*. Another option is to double-click on the Motor Drive control VI and select the input from the Channel control. Designate your second motor as output channel *CH2*.

d. The motors are designed to receive a value between 1 and -1. The positive numbers are in the forward direction, with 1 configured as the maximum forward speed. The negative numbers correspond to reverse speeds with -1 corresponding to the maximum reverse speed. A value of 0 corresponds to a neutral condition for the motor drive outputs. Connect a constant value of 0.5 to the motor drive connected to channel 1, and a value of -0.5 to channel 2.

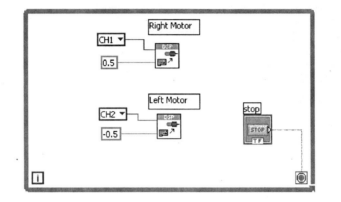

FIGURE 8.11: LabVIEW complete motor test program.

e. The test program is complete.

IV. Running the Motor Activation Program

a. Load the program onto the SPEEDY-33 Board.

b. Now that the program is loaded, connect the external battery source. *Warning*: Make sure the negative terminal of the external battery is connected to the ground pin of the terminal block. The ground pin is closer to the J1 connector. The picture below demonstrates:

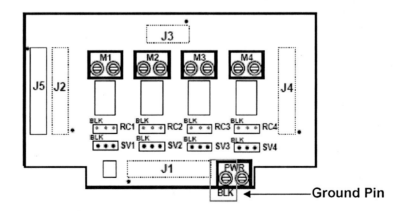

FIGURE 8.12: Robotics daughter module ground pin.

c. On application of the external battery source, the motors should move in opposite directions.

8.5   BUMP SENSOR DETECTION PROGRAM

I.   Build Bump Sensor Detection

   a. Load the bump sensor test program created in STEP I.

   b. Delete one of the bump sensor units in the program, and change the input of the remaining bump sensor to a control instead of a constant.

   c. Place a *Less Than Symbol* into the program by selecting *Functions* → *Comparison* → *Less?* Connect the output of the Bump sensor to the top input *(x)* of the less than symbol. On the bottom input, place a constant value of 0. The *less than* (<) symbol will then output the Boolean value True when the bump sensor is activated and a false value when it is deactivated.

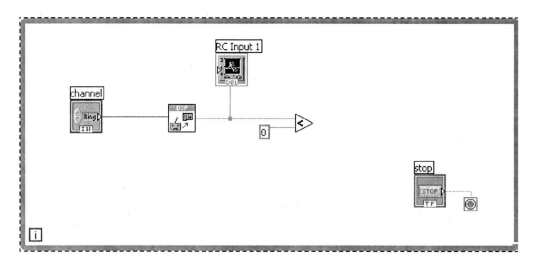

FIGURE 8.13: LabVIEW bump sensor detection program.

   d. Enter the *Functions* → *Elemental I/O* → *Digital Input* menu to introduce a digital switch into the program.

   e. Double-click on the Digital Input Switch and rename the switch "Motor ON/OFF".

   f. Click the *General tab* and change the *Resource* to *on-board DIP switch*. Click the *Configuration tab* and select a digital input switch number between three and eight. Remember DIP switches one and two are off limits for RC1 and RC2.

   g. Insert an *AND Gate* into the program. Connect the Digital Input Switch and the output of the LESS than symbol to the inputs of the AND Gate.

Insert a Boolean Indicator to the output of the AND gate. This gate will activate the Output LED, only when the bump sensor is activated and the Motor ON/OFF digital switch is on. The completed program is shown below.

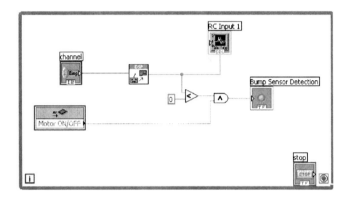

FIGURE 8.14: LabVIEW complete bump sensor test program.

II. Test Bump Sensor Detection Program
 a. Load the program onto the SPEEDY-33 board.
 b. Do not activate the digital input switch and press the bump sensor. The output LED should remain deactivated.
 c. Activate the digital input switch. When the bump sensor is tripped the output LED should illuminate.

III. Saving Test Program as Sub-VI
 a. Create a SUB VI by selecting the following section of code and opening *Create sub-VI* from the *Edit Menu.*

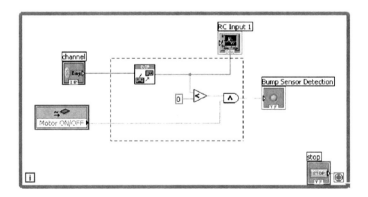

FIGURE 8.15: LabVIEW sub-VI.

b.   Click on the sub-VI, displayed below, and rename the controls and indicators for easier connection during development of the Basic Robot Program.

Rc Input Channel ———— Output for Graph
Motor ON/OFF ·········· Bump Status
Bump_Sensor_Detect_sub.vi

FIGURE 8.16: LabVIEW renaming Sub-VI.

c.   Save the sub-VI.

8.6   BASIC ROBOTIC PROGRAM

I.   Overview

The basic Robotic program should allow a robot to autonomously maneuver through a course, with the bump sensors acting as a guidance system. The robot should move forward until one of the bump sensors are activated. When this occurs, the robot should turn away from the object. The flowchart below describes the algorithmic process the program will follow:

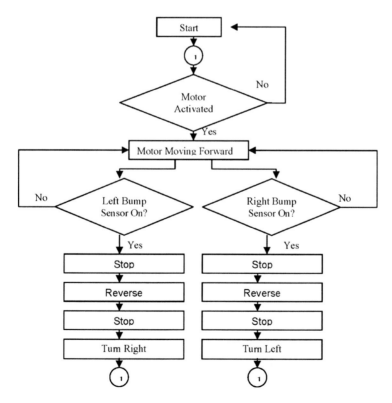

FIGURE 8.17: Robot program flow chart.

II. Build the Test Program

    a. Open a blank VI, and enclose the program in a while loop.

    b. Insert a Digital Input Switch named *Motor ON/OFF*. Remember, do not configure the input to DIP switch 1 or 2 for proper operation of bump sensors.

    c. Add a Select Comparator located in the *Functions → Comparison* menu. This function examines the value of a Boolean input and, from that value, passes one of the two designated inputs to the output. The high input, designated *t*, is sent to the output when the Boolean input is true. The lower input, *f*, is sent to the output when the Boolean input is false.

    d. Connect the *Motor ON/OFF* digital input switch to the Boolean input of the select comparator and place a constant value of 0.5 to the top input and a value of 0 to the bottom input of the select comparator. Therefore, the motors will only activate when the *Motor ON/OFF* switch is on.

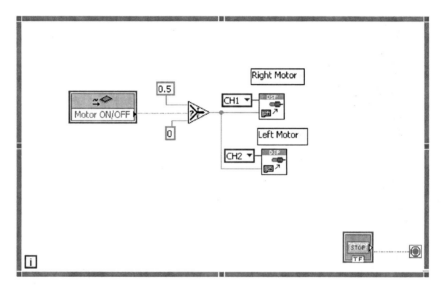

FIGURE 8.18: LabVIEW motor on/off switch.

    e. Insert two instances of the sub-VI created in the last section, and connect the inputs as shown. *Note*: Access to the RC Inputs starts with 0, instead of 1.

    f. Connect a waveform chart to the graph outputs of the sub-VIs. Connect the bump status output of each sub-VI to the selector of a Case structure. The *Case Structure* can be located in the *Functions → Structures* Palette. This is a conditional structure that performs a subdiagram determined by the Boolean input, which in this case is the activation of a bump sensor.

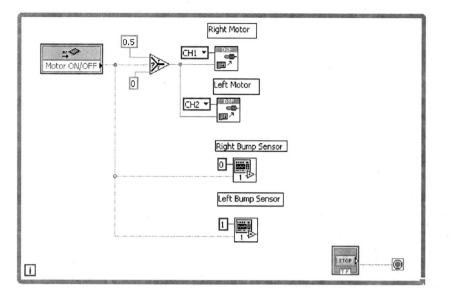

FIGURE 8.19: LabVIEW connecting sub-VI.

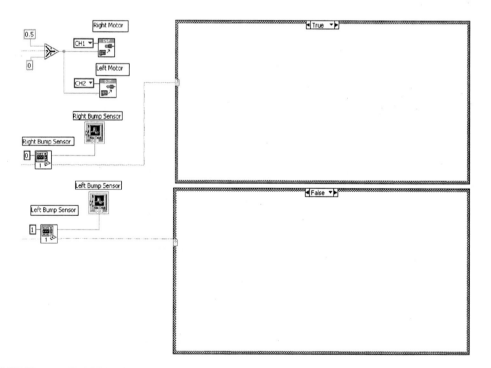

FIGURE 8.20: LabVIEW connecting output waveforms.

g. We will begin working with the True diagram of the Case Structure for the right bump sensor. This diagram occurs after the right bump sensor has been activated. Remember from our flowchart diagram that when this action occurs the motors are to stop, reverse, stop again, turn left, stop, and then continue moving forward. To perform this algorithmic process, a *Stacked Sequence Structure* is added inside the Case structure from the *Functions* → *Structures* Palette.

h. Right-click on the Stacked Structure and select *Add Frame After*. Our algorithm requires five steps, so add four more frames.

i. In the first frame, *frame 0*, we need to turn the motors off for a specific period of time. Select a *For loop* from the *Functions* → *Structures* Palette. A For loop performs the code in its structure a specified number of times. Right-click on the *N*, located in the upper left-hand corner, and place a constant value of 125,000. This is enough time for the first frame. *Note*: The timing for each of your robot will fluctuate; do not except this number as a hard limit.

j. Inside the For loop, add the left and right Motor Drive Control and place their value to 0.

k. Add a For loop to the other three frames, and place the following values displayed in the pictures below. In frames 1 and 3, increase the iteration times to 300,000. In the final frame, when the robot has to turn left away from the object, the right motor should be moving forward, while the left motor should be moving backward.

l. Repeat the same procedure for the left bump sensor, changing only rotation of the motors in the last frame. This is done, so the robotic turns right, instead of left.

Frames 0, 2 and 4

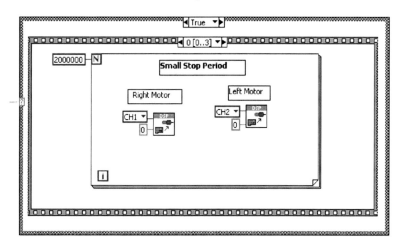

FIGURE 8.21: LabVIEW stop period.

Frame 1

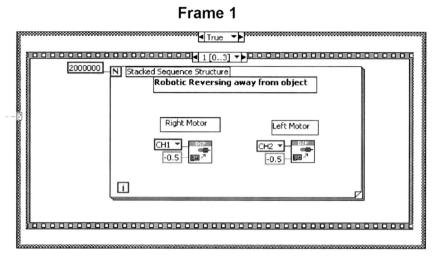

FIGURE 8.22: LabVIEW reverse period.

Frame 3

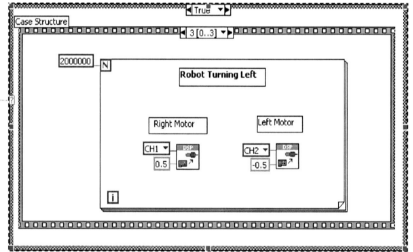

FIGURE 8.23: LabVIEW turn period.

m. Replace the stop button for the while loop with a Digital Input Switch named *Program Termination*.

III. Testing Basic Robot Program

a. Run the program on the SPEEDY-33 Board. Initially, nothing should happen.

b. Turn on the Motor ON/OFF Digital Input Switch, and the motors should move forward.

c. Press the left Bump Sensor. The motors should stop, reverse, stop, turn right (Right Motor reverse and Left Motor forward), stop, and then continue forward.

d. Press the right Bump sensor. The motors should stop, reverse, stop, turn left (Right Motor forward and Left Motor reverse), stop, and then continue forward.

e. Press the Program Termination switch and the program ends. If the program is in one of the bump sensor routines, it will end when it returns from the current routine.

8.7   COUNTER PROGRAM

I. Overview

The counter program allows the robot to maneuver itself from corners. Currently, if the robot meets a corner, the basic program will simply bounce back and forth between the two sides. This new program will perform a 180° turn if the program enters a bump sensor sequence of right, left, right, left. The flowchart below describes the steps that will be taken:

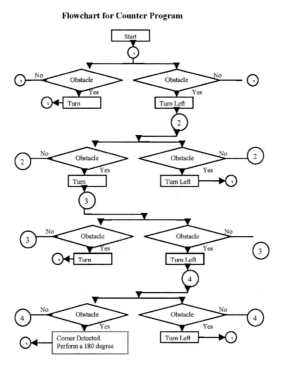

FIGURE 8.24: Counter program flow chart.

II. Building the Counter Program

a. Save a new copy of the Basic Robot VI, and create sub-VIs of the two case structures as shown below.

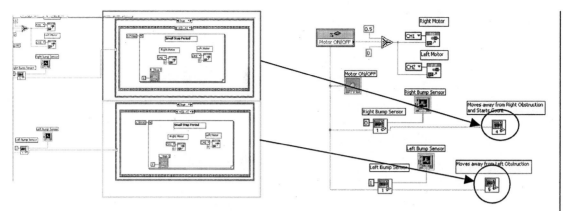

FIGURE 8.25: LabVIEW new sub-VIs.

b. Save the sub-VIs. Enter the sub-VI that responds to a Right Sensor Detection by turning left.

c. In this sub-VI, create a fifth frame and insert a while loop.

d. In this while loop, the motors need to move forward until an obstacle is detected. If the obstacle is on the right side, the sub-VI should terminate to the Main VI because two right bump sensors have been detected in a row. However, if a left bump sensor is detected, the program should stay in the sub-VI and turn right. The picture below shows

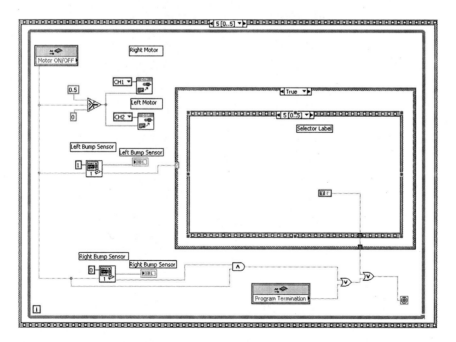

FIGURE 8.26: LabVIEW counter loop.

the content of the while loop. The only differences in this loop is if the right bump is activated the loop terminates. In addition, a frame will be added at the end to terminate the loop and return to the Main VI.

e. Now open a blank VI, and paste the newly constructed while loop inside. Swap the location of the right and left bump sensors, and the direction of the motors in the third frame, so they turn left, instead of right. Save this new VI as Count_2.

f. Return to the Right Bump Sensor Detection VI. In the while loop add a new frame between Frames 4 and 5. In this frame, paste the VI you just created.

g. To view the current progress of your program, select *Show VI Hierarchy* from the *Browse Menu*. This window shows the graphical representation of the embedded Virtual Instruments for each VI and its sub-VIs. At this point, the graphical hierarchy should look similar to this.

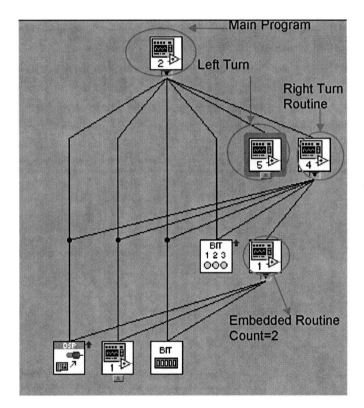

FIGURE 8.27: LabVIEW sub-VI hierarchy.

h. The last step is to insert the sub-VI that will turn the robot 180° if a left bump sensor is pressed. Open a blank VI, and copy the contents of Count_2 into the document. Exchange the left and right bump sensor position. In addition, extend the robot turning time in Frame 3 by a factor of two to 600,000. Therefore, the robot will turn completely around. Save this VI as Count_3.

i. Create a new frame between Frames 4 and 5 of the Count_2 VI. In this new frame, add the newly created Count_3 VI.

j. The program is finished, and the Hierarchy should look similar to the following diagram:

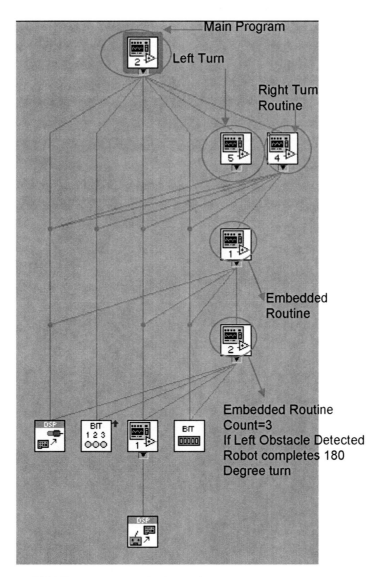

FIGURE 8.28: LabVIEW complete sub-VI hierarchy.

k. The program should perform a 180° turn if the following sequences of bump sensor activations occur: right, left, right, left. Otherwise, the program will behave like the basic robot program.

REFERENCES

1. Hyperception, *Robot Daughter Module for SPEEDY-33™ DSP Board*, http://www.hyperception.com/pdf/manual/us/robot_daughter_card_manual.pdf.
2. Jeff McDougall, *Texas A&M University Implements NI LabVIEW DSP in Curriculum*, National Instruments, 12 February 2006, http://sine.ni.com/csol/cds/item/vw/p/id/619/nid/124200.
3. Marc P. Christensen, Scott C. Douglas, Sally L. Wood, Christopher Kitts, and Ted Mahler, "The Infinity Project Brings DSP Brains to Robots in the Classroom," in *IEEE 11th Digital Signal Processing Workshop & IEEE Signal Processing Education Workshop*, 2004, pp. 88–91.
4. Kevin Springer, Ahmed Saheb, Corby White, and Daniel Akst, *West Texas BEST—Robot Project*, Texas Tech University, 26 April 2006, http://www.ee.ttu.edu/robot/EE3332001G2_Final.ppt.
5. Cory Blair, Michael Castillo, and Sterling Thomas, "Introduction to Robotics," *Senior Design Project Report* (Supervisor: Prof. Lina Karam), Arizona State University, 2006.

* * * *

CHAPTER 9

AM Radio

9.1   OVERVIEW

Amplitude modulation (AM), still widely used in commercial radio today, is one of the simplest ways that a message signal can modulate a sinusoidal carrier wave. The purpose of this chapter is for you to gain familiarity with the concepts of amplitude modulation and demodulation.

9.2   BACKGROUND

9.2.1 AM Modulation/Demodulation

All analog and many digital communication systems that transmit data over cable or using electromagnetic waves use some form of signal modulation. In measurement systems, modulation techniques are useful for measuring very small signals with large amounts of noise.

Consider a signal $V(t) = A \sin\theta(t)$. For a simple sine wave, A is constant, and the phase $\theta(t)$ is a linear function of time, $\theta = wt + \theta_o$, where $w = 2\prod f$ (rad/unit time), and f is cycles/unit time. To use this signal to carry information, we can let either the amplitude A, the frequency w, or the phase θ_o be a function of time. These three possibilities are referred to as amplitude modulation (AM), frequency modulation (FM), and phase modulation (PM). These various types of modulation are often combined; for instance, color television uses AM for brightness, FM for sound, and PM for the hue.

In this chapter, you will study the first of these techniques, amplitude modulation. Consider a signal that you wish to transmit, such as a sound signal $m(t) = m \cos(w_m t)$. To transmit this signal requires you to have a carrier signal, which is typically chosen to be a sinusoid of the form $c(t) = A \cos(wct)$, where wc >> wb (highest frequency in message signal) and the carrier amplitude greater than m, as it makes demodulation easier.

The AM modulator block is shown in Figure 9.1.

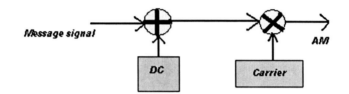

FIGURE 9.1: AM modulator.

There are several ways to demodulate an AM signal. The one that will be used here is using a *product detector*. The way this works is simply by taking the AM signal, multiplying it with the carrier, then applying a lowpass filter to retrieve the transmitted message signal.

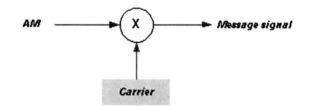

FIGURE 9.2: AM demodulator.

9.3   EXPERIMENT

The experiment is divided in two main parts. The first will consist of the design and simulation of an AM transmitter using LabVIEW and the downloading and testing of the designed system on the Speedy-33; the second part is the demodulation of the AM signal.

9.3.1   AM Transmitter

The AM transmitter will be simulated in LabVIEW using the block diagram for the AM modulator in Figure 9.1. Your AM Modulator should look like Figure 9.1.

The message to be modulated will be simulated using the Frequency Sweep Generator

found in the Embedded Signal Generation subpalette.

Use amplitude of 1 V and sweep the frequency from 100 to 700 Hz, by double-clicking on the Frequency Sweep Generator and configuring the Amplitude and Frequency labels as shown in Figure 9.3.

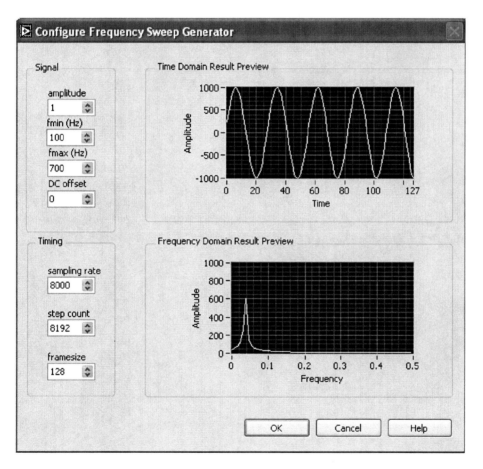

FIGURE 9.3: Configure Frequency Sweep Generator.

As shown in the block diagram previously, the input signal must be added to a DC component, and this combined signal will be multiplied by the carrier frequency. Use a DC component of 1. The carrier frequency is generated by another discrete-time sine wave generator, with amplitude of 32,000 and 16 kHz. Add the appropriate blocks to obtain the diagram shown in Figure 9.4:

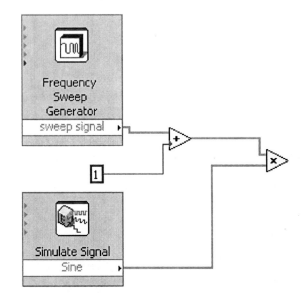

FIGURE 9.4: AM modulator.

Connect the corresponding blocks to view the input and output signals in time and frequency domains. The Block Diagram and Front Panel should look as shown in Figures 9.5 and 9.6, respectively:

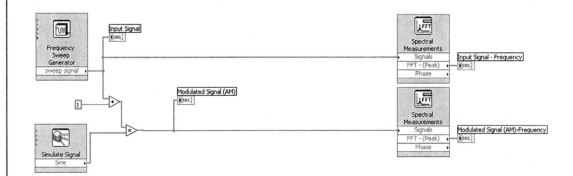

FIGURE 9.5: Final block diagram.

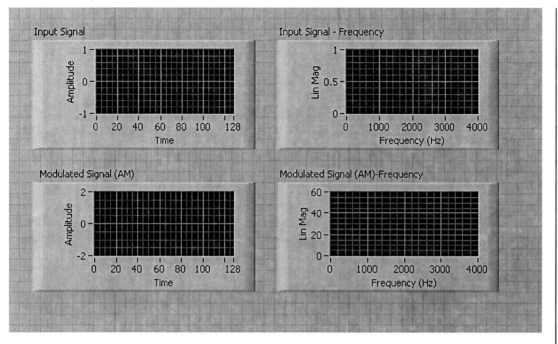

FIGURE 9.6: Final control panel.

The last step is to transmit the AM-modulated signal through an antenna. The speaker on the SPEEDY-33 board acts as the "transmitter antenna". Therefore, connect the AM modulate signal to the Analog Output block as shown in Figure 9.7:

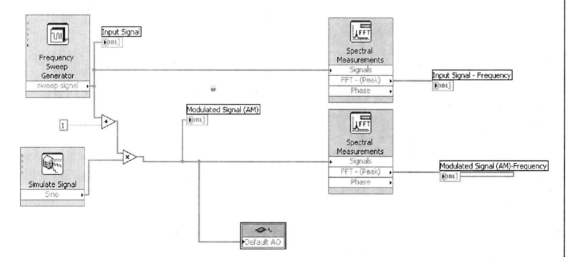

FIGURE 9.7: Transmitting AM signal through Analog Output.

Run your simulation, and observe the signals in time and frequency domain.

Describe the signals in the time domain. You should observe an AM signal.

Observe the frequency domain plot that your simulation should produce. Explain both the magnitude and frequency values.

Connect headphones to the speaker line of the SPEEDY-33 board. Describe what you hear.

9.3.2   AM Receiver

The modulator should run fine, but the intention is to transmit the desired signal and receive it after it goes through modulator, amplifiers, antenna, channel, and receiving antenna.

The received signal must be demodulated to extract the modulating signal, which is the original signal containing the information. In our case, this information is the signal generated from the frequency sweep generator.

The AM receiver that is going to be implemented is just like any AM radio, but it receives signal using the NI-SPEEDY-33 microphone instead of an antenna. This VI demodulates the modulated music that is "broadcasted" by the AM Transmitter implemented earlier. To demodulate the AM signal, one should simply multiply the received signal by the carrier sinusoidal signal as shown in Figure 9.8:

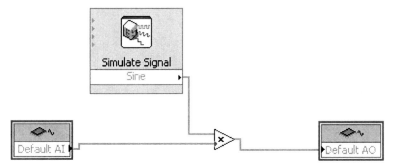

FIGURE 9.8: AM demodulator.

[Add a lowpass filter or mention that Default AO has an integrated lowpass filter.]

Connect the second NI-Speedy-33 board's line output to a speaker, and make sure the board's Input Level jumper is set to Mic input. Run the VI.

Place this board's microphone close to the speaker of the first board that is running as a transmitter.

Connect a graph to the incoming signal and the output of the receiver. Did it recover the modulating signal?

CHAPTER 10

Modem

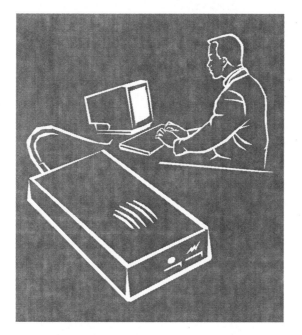

10.1 OVERVIEW

This chapter introduces Frequency Shift Keying (FSK), which is used in modems and which is the process of frequency modulation in which the modulating signals shifts the output frequency between two predetermined frequency values. The DSP target produces an FSK signal by generating, through the DAC output channel on the DSP target, a 400-Hz sinusoid to represent a zero or a 700-Hz sinusoid to represent a one.

10.2 BACKGROUND
10.2.1 What Is a Modem?

Modem is an acronym for Modulator Demodulator. A modem is a device that converts data from digital computer signals to analog signals that can be sent over a phone line. This is called modulation. The analog signals are then converted back into digital data by the receiving modem. This is called demodulation.

A modem is fed digital information, in the form of ones and zeros, from the CPU. The modem then analyzes this information and converts it to analog signals, which can be sent over a phone line. Another modem then receives these signals, converts them back into digital data, and sends the data to the receiving CPU.

"Analog" refers to information being presented continuously, while "digital" refers to data defined in individual steps. Analog information's advantage is its ability to fully represent a continuous stream of information. Digital data, on the other hand, is less affected by unwanted interference, or noise. In digital computers, data is stored in individual bits, which have a value of either 1 (on) or 0 (off). If graphed, analog signals are shaped as sine waves, while digital signals are square waves. Sound is analog, as it is always changing. Thus, to send information over a phone line, a modem must take the digital data given to it by the computer and convert it into sound, an analog signal. The receiving modem must convert these analog signals back into the original digital data.

The most common modem modulation method for radio communications use is some form of FSK (Frequency Shift Keying). This type of modulation passes nicely through many kinds of radio-based "voice channels".

FSK is a modulation technique used by modems in which two different frequencies in the carrier signal are used to represent the binary states of 0 and 1. Using FSK, a modem converts the binary data from a computer into a binary form in which logic 1 is represented by an analog waveform at a specific frequency, and logic 0 is represented by a wave at a different specific frequency.

10.3   EXPERIMENT

Now that you have an idea about the operation of a modem, you are ready to begin the experiment.

10.3.1 Modem: Modulator

As already mentioned, the data sent to a modem is 0 or 1, as it is digital information. Therefore, the data will be simulated using a random number generator of 0 and 1.

To do that, open LabVIEW DSP Module and choose the execution target to be LabVIEW for Windows.

In the Block Diagram window, insert *Random Number* from the Numeric sub-palette. Switch the execution target to SPEEDY-33 by going to *Operate -> Switch Execution Target* and selecting SPEEDY-33.

The *Random Number* block generates a number between 0 and 1. To generate either 0 or 1 with equal probability, one should compare the output of the *Random Generator* with 0.5. If the generated number is greater than 0.5, then the data to be sent is 1, otherwise it is 0. Add the appropriate blocks so that you obtain the following diagram:

FIGURE 10.1: Random generation of 0 and 1.

The output of the *Greater* block will be Boolean, where *True* represents a 1, and *False* represents a 0. If the data to be sent is 0, then the frequency of the sine wave to be transmitted is 400 Hz. Otherwise, it is 700 Hz. Add the appropriate blocks to obtain the following diagram shown in Figure 10.2:

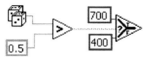

FIGURE 10.2: Frequency selection.

Connect the output of the Select block to the frequency of the sine wave simulated using the EMB sine wave block. The sine wave amplitude should also be controlled by using a slider between 0 and 5,000.

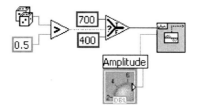

FIGURE 10.3: Sine wave generation.

The final step is to send the sine wave generated to the speaker by connecting the output of the EMB sine wave block to the speaker as shown in Figure 10.4:

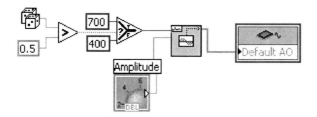

FIGURE 10.4: Analog Output connection.

After adding a While Loop, run the VI and observe the transmitted time domain waveform.

Observe the frequency spectrum of the FSK signal by adding the appropriate graph. Comment.

10.3.2 Modem: Demodulator

The demodulator should convert the analog signal to binary digits depending on the frequency of the sine wave signal.

If the frequency is 400 Hz, then the bit is 0. If it is 700 Hz, then the bit is 1.

The signal is received through the Microphone line, and then, its content at a specified frequency is analyzed using the Goertzel VI found in *Signal Processing > Transforms > Goertzel*. The two frequencies of interest are 400 and 700 Hz. Add the blocks as shown in Figure 10.5:

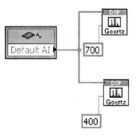

FIGURE 10.5: Frequency analysis using Goertzel VIs.

If the energy of the signal at 700 Hz is higher than the energy at 400 Hz, then the data received is 1. Otherwise, it is 0. Modify the block diagram to obtain the following:

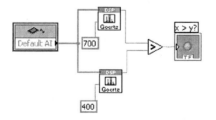

FIGURE 10.6: Converting frequencies to binary digits.

After adding a While Loop, run the demodulator VI simultaneously with the modulator running on a different board, while connecting the Speaker line of the modulator to the Microphone line of the demodulator.

Observe the transmitted time domain waveform by adding the appropriate graph. Compare with the transmitted bits.

Add another graph to observe the frequency spectrum of the demodulated signal by adding the appropriate graph. Comment.

❖   ❖   ❖   ❖

CHAPTER 11

Digital Image Processing Fundamentals

11.1 OVERVIEW

In this chapter, you will explore some of the image management and display functions available as virtual instruments (VIs) in LabVIEW under the NI Vision module.

11.2 BACKGROUND

11.2.1 Digital Image Processing Concepts

For the purposes of image processing, the term *image* refers to a digital image. An image is a function of the light intensity $f(x, y)$ where f is the luminance (brightness) of the point (x, y), and x and y represent the spatial coordinates of a *picture element* (abbreviated *pixel*).

In *digital image processing*, an acquisition device converts an image into a discrete number of pixels. This device assigns a numeric location and *gray-level* value that specifies the brightness of pixels.

The *histogram* of an image indicates the quantitative distribution of pixels per gray-level value. It provides a general description of the appearance of an image and helps identify various components such as the background, objects, and noise.

Histogram Equalization alters the gray-level value of pixels, so they become distributed evenly in the defined grayscale range (0 to 255 for an 8-bit image). The function associates an equal amount of pixels per constant gray-level interval and takes full advantage of the available shades of gray. Use this transformation to increase the contrast of images in which not all gray levels are used.

Thresholding consists of segmenting an image into two regions: a particle region and a background region. In its most simple form, this process works by setting to white all pixels that belong to a gray-level interval, called the *threshold interval*, and setting all other pixels in the image to black. The resulting image is referred to as a *binary image*. For color images, three thresholds must be specified, one for each color component. Thresholding is the most common method of segmenting images into particle regions and background regions. A typical processing procedure would start with filtering or other enhancements to sharpen the boundaries between objects and their background. Then, the objects are separated from the background using thresholding.

11.3   EXPERIMENT

In this experiment, you will learn the basic image processing tools using NI Vision.

11.3.1 Reading an Image File and Image Display

To make use of the large number of image analysis and processing functions available in LabVIEW, an image must be acquired or created. This may be done by using an imaging acquisition process or by simply reading files of images that have already been prepared. We will focus on the latter as well as on using functions for displaying the image.

In LabVIEW, connect the block diagrams shown below to form the VI shown in Figure 11.1.

FIGURE 11.1: Reading an image and displaying it.

In studying the block diagram, activate the LabVIEW "Context Help" and study each VI icon. Try to develop an understanding for the role played by each VI in this simple program.

Run the VI. A dialog will open, and you should select an image file. The program opens the file and displays it.

11.3.2 Selecting Region of Interest

A useful set of functions available in NI Vision are those associated with defining "regions of interest" or ROI. Clearly, it is necessary to have means for identifying where to focus a particular processing step, or marking area(s) of an image to extract, and these tools are commonly used in applications such as machine vision.

You can find VIs for ROIs under the function palette Vision Utilities in the submenus: External Display VIs, Region of Interest VIs, or Region of Interest Conversion VIs. There are also ROI functions under Machine Vision in the submenu Region of Interest VIs. This last one has some simple ROIs we can demonstrate easily.

There are many ways to use ROIs, but one way of immediate concern is to define a selected area for processing. Under the *Machine Vision* functions, a ROI can be selected by: (a) point, (b) line, (c) rectangle, or (d) annulus. Consider just "c" for this exercise, which is highlighted below.

For each gray level in the image [from 0 (black) to 255(white)] we count the number of pixels colored in this gray level. The results of this counting process are displayed as a graph of number of pixels versus gray levels. This graph is called a histogram of the image.

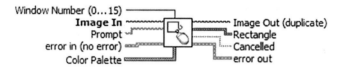

IMAQ Select Rectangle

Allows the user to specify a rectangular area in the image. IMAQ Select Rectangle displays the image in the specified window and provides the rectangle and rotated rectangle tools. IMAQ Select Rectangle returns the coordinates of the rectangle selected when the user clicks OK in the window.

In a basic application, only the "Image In" and "Error In" need to be provided, and the output "Rectangle" can then be sent to other IMAQ Machine Vision functions. If the Image Out is sent to a display, the rectangle area will be displayed as well.

Use the IMAQ Light Meter (Rectangle) function. (Look under Machine Vision->Measuring Intensities VIs). The help for this VI is shown below:

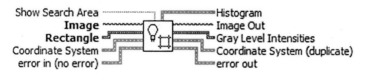

IMAQ Light Meter (Rectangle)

Measures the pixel intensities in a rectangle of an image.

This VI requires the Image information, the defined Rectangle area information, and the error in. The outputs of interest are the gray level intensities (telling you the pixel intensities in the selected region, including their statistics) and the distribution in terms of a histogram. Figure 11.2 shows a complete VI:

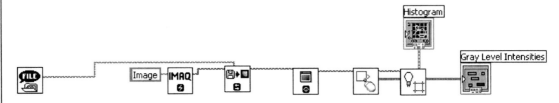

FIGURE 11.2: Selecting a Region of Interest.

11.4   HISTOGRAM EQUALIZATION

Histogram equalization is one of the most important part of the software for any image processing. It improves contrast, and the goal of histogram equalization is to obtain a uniform histogram. This technique can be used on a whole image or just on a part of an image.

Histogram equalization works best on images with fine details in darker regions. Some people perform histogram equalization on all images before attempting other processing operations. This is not a good practice, as good quality images can be degraded by histogram equalization. With a good judgment, histogram equalization can be a powerful tool.

In LabVIEW, use the following blocks to create a VI that displays the original image, a plot of its histogram, the enhanced image, and a plot of its histogram.

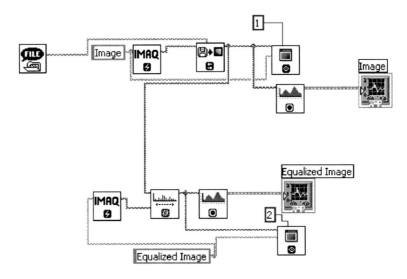

FIGURE 11.3: Histogram Equalization.

Run the VI and notice the difference between the original image and the enhanced one. Comment on the contrast.

11.5   THRESHOLDING

In many vision applications, it is useful to be able to separate out the regions of the image corresponding to objects in which we are interested, from the regions of the image that correspond to background. Thresholding often provides an easy and convenient way to perform this segmentation on the basis of the different intensities or colors in the foreground and background regions of an image.

In addition, it is often useful to be able to see what areas of an image consist of pixels whose values lie within a specified range, or *band* of intensities (or colors). Thresholding can be used for this as well.

Using the blocks shown below, create the VI in Figure 11.4.

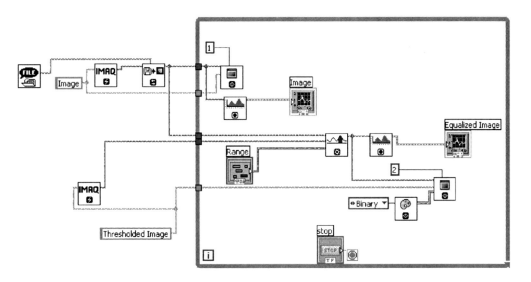

FIGURE 11.4: Thesholding.

Run the VI and notice the thresholded image for different threshold values.

CHAPTER 12

Applications Using USB Camera

12.1   OVERVIEW

In Chapter 11, the main goal was to explore the basics of image management and analysis using LabVIEW Vision. In this experiment, we examine a way to acquire and process images or a video sequence in real-time using a low-cost USB camera.

12.2   BACKGROUND

12.2.1 Acquiring Image Using USB Camera

In this experiment, we examine a way to acquire images using a low-cost USB camera. There are pros and cons to doing this. First, the advantages are: (a) these solutions can be used by anyone using low-cost hardware along with NI LabVIEW software (available to many university students through site license), (b) we can take advantage of relatively new VIs that simplify implementation, and (c) these solutions are highly mobile, as USB cameras can be used on most laptop computers.

The disadvantages include: (a) USB cameras (although adequate for the present purposes) generate relatively low-resolution images and (b) the image acquisition via USB may be too slow for some applications. Older "webcam" drivers had problems with cameras that were not compatible, such as very low cost CMOS type cameras. However, the new USB drivers from NI work with all cameras that have been tested (including Lego cameras, for example).

There is an extensive knowledgebase at the National Instruments website on vision applications, camera selection, etc. Any serious application in vision should take into account a wider survey of hardware capabilities, especially for those applications requiring high resolution, high throughput (large data transfers, which are enabled by image acquisition boards), etc.

The main goal of this chapter is to gain some experience with building a VI that includes image acquisition hardware, software, and utilization of basic machine vision routines found in the NI Vision function library.

12.3   EXPERIMENT

In this experiment, you will learn how to acquire images using a low-cost USB camera and process them.

12.3.1 Checking the Camera Operation

It is assumed that a camera and its driver software have been installed. To check its functionality, double-click on "My Computer" in Windows and look for the camera icon. Open the camera and a fully functional camera view should open.

This mode is useful for quickly setting up the camera for subsequent steps. We will not be conducting any detailed calibration in this exercise, so it is not necessary to spend significant time trying to align the camera.

Close the window after testing the camera.

12.3.2 Testing the Camera With LabVIEW IMAQ USB Vision

The status of available USB cameras can be checked directly using LabVIEW IMAQ Vision functions. This is accomplished using new IMAQ USB Vision functions made available by National Instruments in Spring 2005.

The *IMAQ USB* VIs can be found by following the menu sequence to access the palette:

NI Measurements -> Vision -> IMAQ USB

The menu palette will appear as shown in Figure 12.1:

FIGURE 12.1: Menu palette.

To demonstrate how LabVIEW can access available cameras, open a blank VI. In the block diagram, drop an *IMAQ Enumerate Cameras.vi*. A description of this VI is given below:

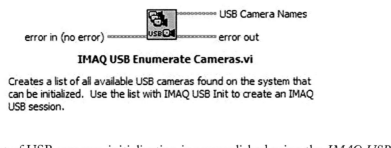

IMAQ USB Enumerate Cameras.vi

Creates a list of all available USB cameras found on the system that can be initialized. Use the list with IMAQ USB Init to create an IMAQ USB session.

Given the list of USB cameras, initialization is accomplished using the *IMAQ USB Init.vi*. Any camera listed in the *string array* (USB Camera Names above) can be selected by its corresponding index (0, 1, etc.).

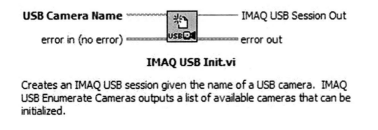

IMAQ USB Init.vi

Creates an IMAQ USB session given the name of a USB camera. IMAQ USB Enumerate Cameras outputs a list of available cameras that can be initialized.

Once selected, the initialization allows any camera to be referenced by IMAQ USB functions. This process is demonstrated next.

If you are putting together a system with more than one USB camera, you can only acquire from one camera at a time. However, as the USB Enumerate VI builds an array of the available cameras, any of these can be selected as needed. For the single camera case, the USB Camera Names array will contain the name at index 0.

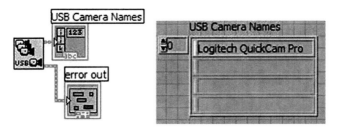

FIGURE 12.2: USB camera name.

The new IMAQ USB VIs have been tested and found to work with several camera models. Those tested include: (a) a Lego USB camera—CMOS, (b) Logitech QuickCam (messenger)—

CMOS, and (c) Logitech QuickCam Pro 4000—CCD. Also, the image acquisition VIs provide image type data that can be analyzed by IMAQ functions directly.

12.4   SIMPLE USB CAMERA IMAGE ACQUISITION

Let us begin by testing the use of the USB camera with the LabVIEW software.

IMAQ Create

Creates an image.

IMAQ USB Snap.vi

Performs a single shot acquisition. Only one camera can acquire at a time.
Use the New Image output from IMAQ Create.vi for the Image in.

IMAQ USB Close.vi

Closes a session to a USB camera that was opened with
IMAQ USB Init.

IMAQ Dispose

Destroys an image and frees the space it occupied in
memory. This VI is required for each image created in an
application to free the memory allocated to IMAQ Create.
Execute IMAQ Dispose only when the image is no longer
needed in your application. You can use IMAQ Dispose
for each call to IMAQ Create or just once for all images
created with IMAQ Create.

IMAQ USB Init.vi

Creates an IMAQ USB session given the name of a USB camera. IMAQ
USB Enumerate Cameras outputs a list of available cameras that can be
initialized. The Video Mode paramater allows you to specify a particular
acquisition mode of the camera. Use the IMAQ USB Property Page.vi
with Video Mode to see the available Video Output formats the camera
supports.

Using the blocks shown above, create the VI shown in Figure 12.3.

FIGURE 12.3: Snapshot acquisition.

Run the VI. It should display a "snapshot" of an image exposed to the lens.

Make sure the camera is not being used by Windows.

Study the block diagram. Note that after the initialization, an IMAQ Create VI is needed to create an image. The "Snap" and "Close" VIs are then used to acquire the image and close the device.

It is important to point out why the device and image are closed and disposed of, respectively, as shown. Closing the device allows others to be opened and frees up resources. Closing the image also makes resources available. Also, LabVIEW may experience problems if a VI is "stopped" before the device is closed by IMAQ USB Close.

12.5 USB CAMERA VIDEO ACQUISITION

On the front panel, create an image display (go to Vision -> Image Display). Make sure to right-click on the display and select "Snapshot" so the image will be retained for viewing (a check mark will appear in the floating menu by "Snapshot" if it is selected properly).

With the addition of a loop structure, you can continuously "grab" images and process them, or display the camera output continuously.

Make the changes shown below using the IMAQ USB Grab Setup and IMAQ USB Grab Acquire VIs.

IMAQ USB Grab Setup.vi

Starts a continuous acquisition. Once the acquisition has started, call IMAQ USB Grab Acquire to copy images from the continuous acquisition. Only one camera can acquire at a time.

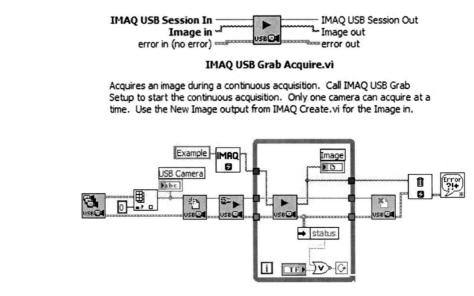

IMAQ USB Grab Acquire.vi

Acquires an image during a continuous acquisition. Call IMAQ USB Grab
Setup to start the continuous acquisition. Only one camera can acquire at a
time. Use the New Image output from IMAQ Create.vi for the Image in.

FIGURE 12.4: Video acquisition.

You should experiment with these basic VIs and with LabVIEW until you see that by combining what we have collected here for "image acquisition" along with the basic "image management" VIs from Experiment 10 you can now proceed to some basic, yet very useful, machine vision exercises.

12.6   BASIC MACHINE VISION WITH USB CAMERA

In this final step, we will introduce a simple machine vision VI to demonstrate the capability of this image acquisition and analysis system.

Recall that in Experiment 10, a method was introduced for selecting an area in an image using a rectangle. That IMAQ VI is summarized again below.

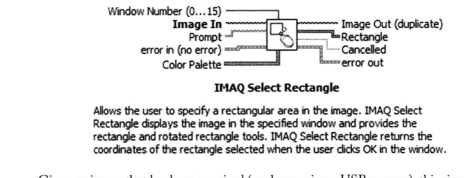

IMAQ Select Rectangle

Allows the user to specify a rectangular area in the image. IMAQ Select
Rectangle displays the image in the specified window and provides the
rectangle and rotated rectangle tools. IMAQ Select Rectangle returns the
coordinates of the rectangle selected when the user clicks OK in the window.

Given an image that has been acquired (as above using a USB camera), this simple area selection method can be used to define the "Rectangle" information required by other vision analysis VIs. Consider, for example, the IMAQ Count Objects VI summarized below:

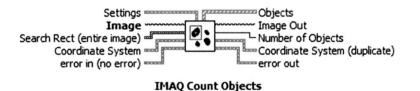

IMAQ Count Objects

Locates, counts, and measures objects in a rectangular search area. This VI uses a threshold on the pixel intensities to segment the objects from their background.

This VI will locate, count, and measure objects in a rectangular search area. A simple VI that allows a user to acquire an image, select a rectangular area on that image, and then have that area analyzed with the count objects VI is illustrated below.

NOTE: In Experiment 10, it was not specified that the "count objects" VI requires a grayscale image. The IMAQ Cast Image VI:

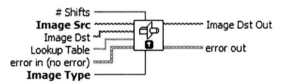

IMAQ Cast Image

Converts the current image type to the image type specified by Image Type. If you specify a lookup table, IMAQ Cast Image converts the image using a lookup table. If converting from a 16-bit image to an 8-bit image, the VI executes this conversion by shifting the 16-bit pixel values to the right by the specified number of shift operations and then truncating to get an 8-bit value.

will allow conversion to Grayscale (U8) as shown below. Also, note that two images are displayed by this VI. The one displayed after the processing by "Count Objects" is used to display the object centers and locations. It is necessary to use the "Settings" and "Objects" parameters as explained in Figure 12.5.

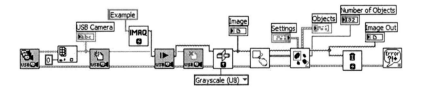

FIGURE 12.5: Counting objects.

It is important to use the "Settings" control for the IMAQ Count Objects VI, as this will allow you to adjust different features of the algorithm (threshold, etc.).

The outputs "Objects" and "Number of Objects" should be used together. The Objects is a cluster of information indexed by the object number. The object information includes location, size, etc.

The relevance of this very useful function to practical applications in machine vision is quite evident. The extensive library of machine vision VIs available in NI Vision should be reviewed to gain full appreciation for the capability of these tools.

*   *   *   *

APPENDIX A

VIs at a Glance

Add

Computes the sum of the inputs.

Analog Input

Reads data from the analog input you specify in the **Configure Elemental I/O** dialog box. Configure the Analog Input function to read the analog input signals connected to a LabVIEW Embedded target. Double-click the Analog Input function on the block diagram to display the **Configure Elemental I/O** dialog box.

Analog Output

Writes data to the analog output you specify in the **Configure Elemental I/O** dialog box. Configure the Analog Output function to read the analog input signals connected to a LabVIEW Embedded target. Double-click the Analog Output function on the block diagram to display the **Configure Elemental I/O** dialog box.

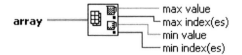

Array Max & Min

Returns the maximum and minimum values found in **array**, along with the indexes for each value.

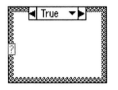

Case Structure

Has one or more subdiagrams, or cases, exactly one of which executes when the structure executes. The value wired to the selector terminal determines which case to execute and can be Boolean, string, integer, or enumerated type. Right-click the structure border to add or delete cases. Use the Labeling tool to enter value(s) in the case selector label and configure the value(s) handled by each case.

Digital Input

Reads data from the digital input you specify in the **Configure Elemental I/O** dialog box. Configure the Digital Input function to read the digital input signals connected to a LabVIEW Embedded target. Double-click the Digital Input function on the block diagram to display the **Configure Elemental I/O** dialog box.

Digital Output

Writes data to the digital bank output you specify in the **Configure Elemental I/O** dialog box. Configure the Digital Bank Output function to write the digital bank output signals connected to a LabVIEW Embedded device. Double-click the Digital Bank Output function on the block diagram to display the **Configure Elemental I/O** dialog box.

Divide

Computes the quotient of the
inputs.

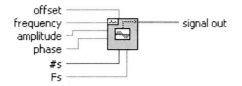

EMB Sine Waveform.vi

Generates a waveform containing a sine wave.

Equal?

Returns TRUE if **x** is equal to **y**.
Otherwise, this function returns
FALSE.

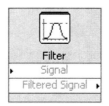

Filter

Processes signals through filters and
windows.

For Loop

Executes its subdiagram n times, where n is the value
wired to the count (**N**) terminal. The iteration (**i**) terminal
provides the current loop iteration count, which ranges
from 0 to n-1.

Frequency Sweep Generator

Generates a sweeping sine waveform.

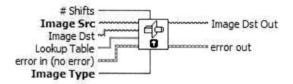

Greater?

Returns TRUE if **x** is greater than
y. Otherwise, this function
returns FALSE.

IMAQ Cast Image

Converts the current image type to the image type specified by
Image Type. If you specify a lookup table, IMAQ Cast Image
converts the image using a lookup table. If converting from a
16-bit image to an 8-bit image, the VI executes this conversion
by shifting the 16-bit pixel values to the right by the specified
number of shift operations and then truncating to get an 8-bit
value.

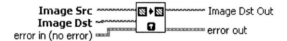

IMAQ Copy

Copies the specifications and pixels of one image into another
image of the same type. You can use this function to keep an
original copy of an image (for example, before processing an
image). The full definition of the source image as well as the
pixel data are copied to the destination image. The border size
of the destination image also is modified to be equal to that of
the source image. If the source image contains additional
information, such as calibration information, overlay information,
or information for pattern matching, this information is also
copied to the destination image.

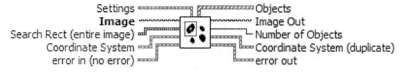

IMAQ Count Objects

Locates, counts, and measures objects in a rectangular search area. This VI uses a threshold on the pixel intensities to segment the objects from their background.

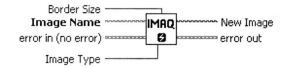

IMAQ Create

Creates an image.

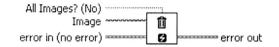

IMAQ Dispose

Destroys an image and frees the space it occupied in memory. This VI is required for each image created in an application to free the memory allocated to IMAQ Create. Execute IMAQ Dispose only when the image is no longer needed in your application. You can use IMAQ Dispose for each call to IMAQ Create or just once for all images created with IMAQ Create.

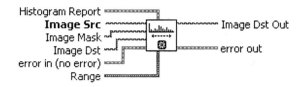

IMAQ Equalize

Produces a histogram equalization of an image. This VI redistributes the pixel values of an image to linearize the accumulated histogram. The precision of the VI is dependent on the histogram precision, which in turn is dependent on the number of classes used in the histogram.

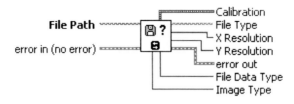

IMAQ GetFileInfo

Obtains information regarding the contents of the file. This information is supplied for standard file formats only: BMP, TIFF, JPEG, PNG, or AIPD.

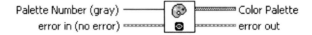

IMAQ GetPalette

Selects a display palette. Five predefined palettes are available. To activate a color palette, choose a code for Palette Number and connect the Color Palette output to the input Color Palette of IMAQ WindDraw.

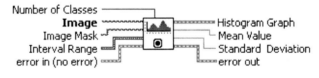

IMAQ Histograph

Calculates the histogram from an image. This VI returns a data type (cluster) compatible with a LabVIEW graph.

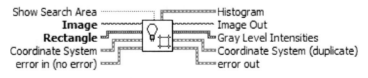

IMAQ Light Meter (Rectangle)

Measures the pixel intensities in a rectangle of an image.

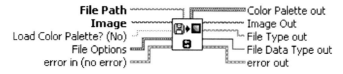

IMAQ ReadFile

Reads an image file. The file format can be a standard format (BMP, TIFF, JPEG, PNG, and AIPD) or a nonstandard format known to the user. In all cases, the read pixels are converted automatically into the image type passed by Image.

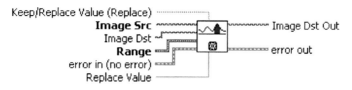

IMAQ Select Rectangle

Allows the user to specify a rectangular area in the image. IMAQ Select Rectangle displays the image in the specified window and provides the rectangle and rotated rectangle tools. IMAQ Select Rectangle returns the coordinates of the rectangle selected when the user clicks OK in the window.

IMAQ Threshold

Applies a threshold to an image.

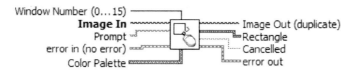

IMAQ USB Grab Acquire.vi

Acquires an image during a continuous acquisition. Call IMAQ USB Grab Setup to start the continuous acquisition. Only one camera can acquire at a time. Use the New Image output from IMAQ Create.vi for the Image in.

IMAQ USB Session In ———— IMAQ USB Session Out
error in (no error) ———— error out

IMAQ USB Grab Setup.vi

Starts a continuous acquisition. Once the acquisition has started, call IMAQ USB Grab Acquire to copy images from the continuous acquisition. Only one camera can acquire at a time.

IMAQ USB Session In

error in (no error) ━━━━━━━━ error out

IMAQ USB Close.vi

Closes a session to a USB camera that was opened with IMAQ USB Init.

━━━━━━━━ USB Camera Names

error in (no error) ━━━━━━━━ error out

IMAQ USB Enumerate Cameras.vi

Creates a list of all available USB cameras found on the system that can be initialized. Use the list with IMAQ USB Init to create an IMAQ USB session.

IMAQ USB Session In ━━━━━━━━ IMAQ USB Session Out

error in (no error) ━━━━━━━━ error out

IMAQ USB Grab Setup.vi

Starts a continuous acquisition. Once the acquisition has started, call IMAQ USB Grab Acquire to copy images from the continuous acquisition. Only one camera can acquire at a time.

USB Camera Name ━━━━━━━━ IMAQ USB Session Out

error in (no error) ━━━━━━━━ error out

IMAQ USB Init.vi

Creates an IMAQ USB session given the name of a USB camera. IMAQ USB Enumerate Cameras outputs a list of available cameras that can be initialized.

IMAQ USB Session In ━━━━━━━━ IMAQ USB Session Out
Image in ━━━━━━━━ Image out
error in (no error) ━━━━━━━━ error out

IMAQ USB Snap.vi

Performs a single shot acquisition. Only one camera can acquire at a time. Use the New Image output from IMAQ Create.vi for the Image in.

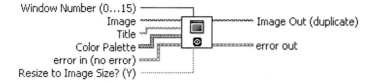

IMAQ WindDraw

displays an image in an image window. The image window appears automatically when the VI is executed. The default image window does not have scrollbars. You can add scrollbars by using the IMAQ WindSetup VI.

Increment

Adds 1 to the input value.

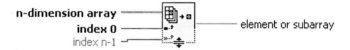

Index Array

Returns the **element or sub-array** of **n-dimension array** at **index**.

Multiply

Returns the product of the inputs.

```
1 2 3
```

Numeric Constant

Use the numeric constant to pass a numeric value to the block diagram. Set this value by clicking inside the constant with the Operating tool and typing a value.

Quotient & Remainder

Computes the integer quotient and the
remainder of the inputs.

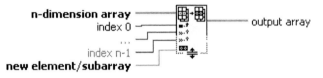

Replace Array Subset

Replaces an element or subarray in an array at the point you
specify in **index**.

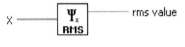

RMS.vi

Computes the root mean square (rms)
of the input sequence **X**.

Select

Returns the value wired to the **t** input or
f input, depending on the value of **s**. If **s**
is TRUE, this function returns the value
wired to **t**. If **s** is FALSE, this function
returns the value wired to **f**.

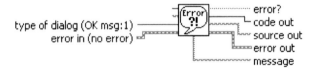

Simple Error Handler.vi

Indicates whether an error occurred. If an error occurred, this VI
returns a description of the error and optionally displays a dialog
box.

Simulate Signal

Simulates a sine wave, square wave, triangle
wave, sawtooth wave, or noise (DC) signal.

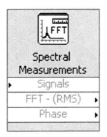

Spectral Measurements

Performs spectral measurements, such as peak spectrum
and power spectrum, on a signal.

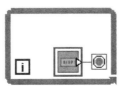

While Loop

Repeats the subdiagram inside it until the conditional
terminal, an input terminal, receives a particular Boolean
value. When you place this While Loop on the block
diagram, a stop button also appears on the block diagram
and is wired to the conditional terminal.

Author Biography

Lina J. Karam received her BS degree in engineering from the American University of Beirut in 1989, and earned her MS and PhD degrees in Electrical Engineering from the Georgia Institute of Technology in 1992 and 1995, respectively. She is currently an associate professor in the Department of Electrical Engineering at the Arizona State University. She is currently the director of the Image, Video, and Usability, the Multi-Dimensional DSP, and the Real-Time Embedded Signal Processing laboratories in the Department of Electrical Engineering at ASU. Her research interests are in the areas of image and video processing, compression, and transmission; human visual perception; multidimensional signal processing; error-resilient source coding; digital filter design; and biomedical imaging. From 1991 to 1995, she was a research assistant in the Graphics, Visualization, and Usability Center and then in the Department of Electrical Engineering at Georgia Tech. She has worked at Schlumberger Well Services (Austin, TX), and in the Signal Processing Department of AT&T Bell Labs (Murray Hill, NJ) in 1992 and 1994, respectively. Prof. Karam is the recipient of a US National Science Foundation CAREER Award. She served as the Chair of the IEEE Communications and Signal Processing Chapters in Phoenix in 1997 and 1998. She was a member of the organizing committees of the 1999 IEEE International Conference on Acoustics, Speech, and Signal Processing (ICASSP99), the 2000 IEEE International Conference on Image Processing (ICIP00), the 2008 IEEE International Conference on Acoustics, Speech, and Signal Processing (ICASSP08), and Asilomar 2008. She is the Technical Program Chair of the 2009 IEEE International Conference on Image Processing (ICIP09). She is an associate editor of the *IEEE Transactions on Image Processing* and is serving on the editorial board of the *Foundations and Trends in Signal Processing Journal*. She also served as an associate editor of the *IEEE Signal Processing Letters* from 2004 to 2006, and as a member of the IEEE Signal Processing Society's Conference Board. She is an elected member of the IEEE Circuits and Systems Society's DSP Technical Committee, and of the IEEE Signal Processing Society's Image and Multidimensional Signal Processing Technical Committee. She is a senior member of the IEEE and a member of the Signal Processing and Circuits and Systems societies of the IEEE.

Naji Mounsef received his BS degree (magna cum laude) in computer and communication engineering from Notre Dame University (NDU), Lebanon, in 2004, and his MS degree in computer and communication engineering from the American University of Beirut (AUB) in 2005. While at NDU, he served as a teaching assistant in circuits, electronics, logic design, and digital signal processing laboratories. From 2004 to 2005, he was also a teaching assistant at the AUB in digital signal processing and digital image processing laboratories, for which he wrote a laboratory manual. During the same period, he worked as a research assistant at AUB and worked on problems related to software radio and efficient turbo decoding. After teaching for a semester at NDU and AUB, he joined the PhD program in Electrical Engineering (2006) at the Arizona State University (ASU), where he first worked as a research assistant and then as a teaching assistant. In summer 2008, he interned at the Translational Genomics Institute. He is part of the Image, Video, and Usability (IVU) laboratory in the Electrical Engineering Department of ASU, and his research areas include image processing and genomic signal processing.

Breinigsville, PA USA
22 February 2010
232994BV00003B/1/P